Hierarchical Operations and Supply Chain Planning

Springer-Verlag London Ltd.

Tan Miller

Hierarchical Operations and Supply Chain Planning

With 65 Figures

Springer

Tan Miller, PhD, M.B.A.
Warner-Lambert Company, 201 Tabor Road, Morris Plains, New Jersey, 07950 USA

British Library Cataloguing in Publication Data
Miller, Tan
Hierarchical operations and supply chain planning
1.Business logistics 2.Strategic planning
I.Title
658.7
ISBN 978-1-4471-1110-8 ISBN 978-1-4471-0305-9 (eBook)
DOI 10.1007/978-1-4471-0305-9

Library of Congress Cataloging-in-Publication Data
A catalog record for this book is available from the Library of Congress.

ISBN 978-1-4471-1110-8

http://www.springer.co.uk

Originally published by Springer-Verlag London Limited in 2002

2nd printing of 1-85233-340-5, published by Springer-Verlag London 2001

Typesetting: Camera ready by author

69/3830-543210 Printed on acid-free paper SPIN 10866270

This book is dedicated to my wife Jeanne, my sons Lucas and Nathan, and my parents who have all supported me and put up with me over the years.

Preface

Hierarchical production planning (HPP) represents an approach and a philosophy towards the organization, planning and scheduling of production activities which has existed both in theory and practice for decades. Originally developed as a planning and scheduling approach primarily for production activities, the philosophy of hierarchical production planning has evolved to encompass a broad range of supply chain management activities ranging from network-wide inventory management and forecasting, to transportation, warehouse planning and scheduling, and so on. In many of these activity areas, understandably no mention or acknowledgment of the hierarchical production planning approach exists. Nevertheless, upon closer inspection, clear similarities in techniques and philosophy exist among the myriad decision-making approaches used today to manage the diverse aspects of the supply chain and those techniques and approaches which embody hierarchical production planning.

The common theme of a hierarchical approach towards the management of decision-making activities across the supply chain suggests that there exist benefits and broad learnings to be gained from exploring the hierarchical approaches and techniques utilized today in managing different components of the supply chain. Similarly, there exist numerous potential benefits and methodological improvements to be attained by applying hierarchical techniques used in certain elements of supply chain planning (e.g., production planning and scheduling) to other supply chain activities. The review of current hierarchical planning approaches and techniques and the exploration of how to more broadly apply these approaches and techniques to supply chain management thus represents a principal objective and contribution of this book.

This book describes the application of hierarchical planning techniques to all the major functional areas of supply chain planning including production, distribution, warehousing, transportation, inventory management, forecasting, and performance management. In particular, the book provides a comprehensive review and understanding of how hierarchical planning techniques and principles can contribute to the effective and efficient management and planning of supply chain activities. The book begins with an introduction to HPP and a discussion of the rationale for using this planning approach in Chapter 1. Then, in Chapter 2, we review some well-known, original hierarchical production planning techniques and implementations dating back several decades. Building on this historical base, in Chapters 3 through 8, we then explore and evaluate numerous more current hierarchical planning methods and applications covering most major supply chain functional activities. In addition to reviewing current hierarchical planning approaches, we introduce and explain new, original hierarchical planning methodologies and algorithms that can be used in different supply chain management capacities. These methodologies and algorithms vary from large-scale conceptual planning frameworks, to simple algebraic formulas, to mathematical optimization models, to equally important "qualitative" approaches such as communication techniques to support hierarchical supply chain operations. Further, many of the methods and techniques offered are based on actual implementation experience and practice in private industry.

This book is written to cover the interests of a wide variety of audiences ranging from private industry practitioners, to academic researchers, to students of operations, logistics and supply chain management and planning. For the practitioner, this book is designed to offer insights on the power of hierarchical planning techniques, as well as "cookbook" type approaches and guidance for practical implementations. For the academic researcher, the book is designed to provide a broad review of hierarchical production and supply chain planning methodologies, as well as selected hierarchical optimization models. Finally, for students of logistics, operations, and supply chain planning and management, this book is designed to provide a broad overview and introduction to planning and scheduling activities that occur across a wide spectrum of functional areas. Thus, this book can also appropriately fill the role of a second "text" to accompany a primary text in an undergraduate or graduate course on logistics, operations management, supply chain or production planning.

Finally, I would like to acknowledge Elaine Grieb who has patiently worked through many iterations of the figures, tables and text found in this book. Without her expertise, it also would not have been possible to complete this project. I would also like to thank Paul Darlington and Joe Masse of Warner-Lambert Company for their interest and support on this project, as well as the many other colleagues at Warner-Lambert with whom I worked in the area of supply chain planning and management. It was an enjoyable and educational experience.

Contents

Chapter 1

Introduction

Hierarchical production planning (HPP) represents an approach and a philosophy towards the organization, planning and scheduling of production activities which has existed both in theory and practice for several decades. Originally developed as a planning and scheduling approach primarily for production activities, the philosophy of hierarchical production planning has evolved to encompass a broad range of supply chain management activities. Today, one can observe what can broadly be viewed as extensions of the HPP philosophy across activities ranging from production planning and integrated distribution and production planning to supply chain wide inventory management, transportation and warehouse planning and scheduling, sales and marketing planning and so on. In many of these activity areas, understandably no mention or acknowledgment of the hierarchical production planning approach exists. Nevertheless, upon closer inspection, clear similarities in techniques and philosophy exist among the myriad decision-making approaches used today to manage the diverse aspects of the supply chain and those techniques and approaches which embody hierarchical production planning. The common theme of a hierarchical approach towards the management of decision-making activities across the supply chain suggests that there exist benefits and broad learnings to be gained from exploring the hierarchical approaches and techniques utilized today in managing different components of the supply chain. Similarly, there exist numerous potential benefits and methodological improvements to be attained by applying hierarchical techniques used in certain elements of supply chain planning (e.g., production planning and scheduling) to other supply chain activities. The review of current hierarchical planning approaches and techniques and the exploration of how to more broadly apply these approaches and techniques to supply chain management represents a principal focus of this book.

1.1 Introduction To Hierarchical Production Planning Systems

The motivation and rationale for employing a "hierarchical" production planning and scheduling process can perhaps best be understood by first considering a general framework for managerial decision-making. In his seminal work on planning and control systems, Robert Anthony (1965) offered a planning framework often quoted and employed by both theoreticians and practitioners in hierarchical production planning. A review of this framework will help to put the hierarchical characteristics of managerial decision making into perspective.

Anthony classified all managerial decisions into three broad categories consisting of: (1) strategic planning, (2) management control and (3) operational control. A number of authors (e.g., Ackoff, 1970) have termed the second category as tactical planning and the third category as operational planning and scheduling. For purposes of this book, we will denote these three categories as strategic planning, tactical planning and operational planning with the understanding that there exist minor variations on these classifications in the literature. We will now examine each of these planning levels in some detail.

1.1.1 Strategic Planning

Strategic planning represents the highest level of the hierarchy of decision making activities which occur within a firm or an organization. Ultimately, strategic planning decisions are concerned with defining the long term objectives of a firm, charting the long term course which will allow a firm to meet its defined objectives, and assuring that a firm has the proper resources and assets necessary to support its long term objectives. In his framework, Anthony (1965) offers the following definition of the role of strategic planning. He states that strategic planning is "the process of deciding on objectives of the organization, on changes in these objectives, on the resources used to obtain these objectives, and on the policies that are to govern the acquisition, use and disposition of these resources". In manufacturing and distribution, there are a number of classic decisions which generally fall into the strategic planning category. Figure 1.1 displays a sample of typical strategic logistics issues and problems which firms must address.

- Plant and warehouse locations, missions and relationships (*i.e.*, network infrastructure and design)
- New plant locations and sizes, and plant closings
- New warehouse locations and sizes, and warehouse closings
- Plant and warehouse capacity levels
- Plant and warehouse technology and equipment acquisition
- Plant and warehouse design
- Mix of owned assets vs. third party resources utilized (*i.e.*, outsourcing decisions)
- Transportation network and transportation providers
- Order fulfillment approach (e.g., make-to-order vs. make-to-stock)

Figure 1.1 Selected Strategic Logistics Decisions

Although Figure 1.1 clearly represents a subset of the universe of strategic logistics issues that firms must tackle, it serves to illustrate the types of decisions which occur at this level of the planning hierarchy. Further, the decisions illustrated in Figure 1.1 share a number of common characteristics found in many strategic production and distribution planning decisions.

First of all, each of the decisions in Figure 1.1 has major implications for a firm and will have a significant effect on the firm for a relatively long time period. This dictates that a firm must evaluate these strategic decisions in the context of a relatively long planning horizon (e.g., a minimum of several years into the future). Typically the level of risk and uncertainty associated with a decision increases as the time period in the future which a decision will impact grows more distant.

Thus, strategic planning decisions generally have a higher degree of risk and uncertainty associated with them than lower level decisions. Strategic planning decisions also frequently involve major capital investments on the part of an organization. Obviously this represents a significant element of the risk of most strategic decisions. As is the case with the decisions in Figure 1.1, frequently strategic logistics decisions require major capital investments (or disinvestments) because they involve adding to or modifying the firm's infrastructure. As these characteristics illustrate, overall, we can say that strategic decisions are extremely important to the ongoing strength and viability of a firm. In the long run, these decisions will determine the competitiveness and profitability of a firm, its long run growth rate, market share and so on. These characteristics of strategic decisions have several implications.

The critical risks, costs, and long lasting effects of strategic decisions dictate that senior managers typically make the majority of these decisions. Inputs may come from all levels of the organization, however, senior management ultimately makes the final call. As observed, strategic decisions are very broad in scope and implications. This typically requires that the information used by managers as inputs to the decision making process provides a broad, well rounded perspective. Thus, the ultimate data inputs to this process must represent aggregated information in order to offer the broad perspective needed by the decision maker at this level. For example, weekly future production schedules at an end item or sku level will provide too much detail for a senior manager attempting to make a plant capacity decision. Strategic decisions also generally require both internal and external information inputs to facilitate effective decision making. For example, managers making long run decisions on planned network-wide manufacturing capacity will certainly want to consider competitors' capacity levels and forecast future total industry demand levels as inputs to their decision. Again, we will observe that this differs from lower level decisions for which internal information inputs are usually sufficient. In summary, the nature of strategic planning decisions generally requires the involvement of senior managers who must employ broad, aggregated informational inputs to make these decisions.

1.1.2 Tactical Planning

Tactical planning represents the second or intermediate level of decision making activities that occur in a firm. Quite often annual planning is viewed as a subset of tactical planning. In Anthony's (1965) framework, the decision making process primarily focuses on resource allocation and resource utilization at this level. Anthony describes that managers' tactical planning activities must obtain and use resources effectively and efficiently to assure the accomplishment of the firm's objectives. This description seems quite intuitive when considered in the context of a planning hierarchy. As just reviewed, strategic planning decisions address such issues as infrastructure and overall capacity levels. Thus, at the next level down, the decision making process must focus on how to most effectively utilize the infrastructure and capacity that the implementation of strategic decisions has created.

There are a number of traditional manufacturing and distribution decisions which frequently occur at the tactical level. Figure 1.2 offers a sampling of some of the more common tactical logistics decisions.

- Assignment of production capacity to product families, by plant (and often) by medium size time periods (e.g., quarters)
- Planned manufacturing capacity utilization rates, by plant and network wide
- Workforce requirements (regular and overtime levels)
- Plant - distribution center - sales region/country sourcing assignments
- Inter-facility (e.g., inter-distribution center) shipment plans
- Inventory investment and deployment plans
- Transportation mode and carrier selections

Figure 1.2 Selected Tactical Logistics Decisions

Figure 1.2 exhibits only a small subset of the multitude of manufacturing and distribution decisions which take place at the tactical planning level. Nevertheless, most of the decisions shown share some common characteristics. First of all, firms evaluate these decisions over a planning horizon which typically spans at least twelve months and sometimes up to eighteen or twenty-four months. (One year probably represents the most common tactical planning horizon.) In industries that experience significant seasonality in demand patterns, the tactical planning horizon should extend at the minimum over the complete seasonal cycle (which is frequently but not always one year). Tactical planning decisions typically do not carry as high a level of risk and uncertainty as do strategic planning decisions. For example, tactical decisions typically do not have major long term implications for a firm's infrastructure (e.g., its network). Nevertheless, decisions at the tactical level often bear significant importance and risk for a firm. For example, annual planning decisions in the retail industry regarding the level of inventory to manufacture or purchase for the Christmas season or the fall and spring fashion seasons can critically impact the fortunes of individual firms. Firms typically make both "expense" and "capital investment" type decisions at the tactical level, although the "capital" type decisions are generally of a smaller scale than strategic "capital investment" decisions. For example, a manufacturer may decide to purchase warehouse or production equipment in its annual planning process. (In contrast, the decision to purchase or build a new plant or distribution center would fall into the strategic planning process.) Thus, as its intermediate level in the planning hierarchy connotes, tactical planning decisions generally influence the operations of a firm over the medium term horizon of about six to twenty-four months at the maximum.

The management level responsible for tactical decision making generally corresponds with the intermediate level that tactical planning holds in the planning hierarchy. Seasoned middle managers and lower level senior executives typically have responsibility for executing the tactical planning process. An organization's most senior management will review or oversee key tactical plans, however, as a rule, the majority of decision-making occurs below the top management level. The twelve to eighteen month planning horizon common at this level dictates that informational inputs provide managers with a relatively broad perspective on the overall business. This necessitates the development of "aggregated" final decision inputs which afford managers this broader perspective. Much of the work done to generate aggregate informational inputs may start at a detailed level (e.g., an inventory analysis may begin at the individual sku level). Ultimately, however, effective tactical planning demands the construction of aggregate decision support inputs. In subsequent chapters, we will observe that a key element of tactical planning consists of developing aggregate inputs which accurately depict important business conditions that exist at a more detailed level. Finally, tactical decisions predominately require internal informational inputs, in contrast to strategic decisions which more heavily rely on external as well as internal information. In summary, the tactical planning process represents a critical element to the long term success of any organization. Experienced managers, using relatively aggregate informational inputs, coupled with good managerial knowledge of the underlying operational processes, bear the brunt of decision making responsibility at this level.

1.1.3 Operational Planning

Operational planning and scheduling represents the third and lowest level of the hierarchical planning process. At this level, the firm must carry out the resource allocation and utilization decisions made at the tactical level in the daily and weekly activities which occur at the operational level. Anthony (1965) described that the responsibilities of managers at this level consist of assuring that the organization performs individual tasks efficiently and effectively, and that these tasks support the higher level tactical plans of the organization. Again this framework seems intuitively appealing from a hierarchical perspective. At the tactical level, the firm makes the resource allocation decisions to facilitate the operations of the business, and at the operational level, the firm executes its daily operations using the resources made available by the tactical planning process.

There exist an extraordinary number of individual manufacturing and distribution decisions that regularly occur at the operational level. Figure 1.3 presents just a small sample of these types of decisions.

Similar to strategic and tactical planning decisions, the operational planning decisions illustrated in Figure 1.3 have a number of fairly common characteristics. First of all, firms make these decisions on a routine basis and the decisions made generally focus on a relatively short planning horizon. Typical planning horizons at this level range from days to several weeks to perhaps the next one or two quarters. Although some operational planning decision support tools (e.g., distribution resource planning and master production scheduling software) may support planning horizons of twelve to eighteen or more months, schedulers using these tools generally concentrate on the near term portion of these schedules. The decisions made at the operational level generally do not have tremendous risk or uncertainty when judged individually. For example, with the exception of certain build to order environments (e.g., jetliner production), the decision of

- **Daily and weekly production scheduling at the item level including item sequencing decisions**
- **Short term inventory balancing and reconciliation**
- **Customer order processing and scheduling**
- **Warehouse operations scheduling**
- **Labor scheduling for manufacturing and warehouse operations**
- **Vehicle scheduling**
- **Carrier selection for individual loads**
- **Logistics support for individual promotions (e.g., special direct store delivery promotions)**

Figure 1.3 Selected Operational Logistics Decisions

whether and how to schedule the delivery of one customer order by the customer's requested delivery date does not hold great risk for a firm. Over the long term, however, the ability of a firm to effectively and efficiently meet customer delivery dates will greatly influence customers' collective views of a firm and the long term viability of a firm. Again though, individual operational decisions have relatively minimal risk compared to strategic and tactical planning decisions. Similarly, individual operational planning decisions also generally do not directly involve significant capital or operating "expense" costs. For example, the daily scheduling of trucks or a production line involves relatively minor cost trade-offs. In the long run, however, inefficient scheduling decisions at the operational level can collectively create substantial unnecessary costs and service failures. One individual scheduling decision on its own, however, rarely involves a major cost trade-off.

The perspective of managers responsible for operational planning in an organization typically matches the role of this level in the planning hierarchy. Colleagues charged with assuring the successful execution of daily operations are the operational planning decision makers. Although essential to an organization's success, employees at this level are generally not considered senior managers. The short run perspective of operational planning both facilitates and requires the use of detailed, disaggregated data inputs in the decision making process. For example, to develop a weekly production schedule sequenced by item for a production line requires that the scheduler review inventory, demand and other relevant factors at the item level. Aggregated product family data will not provide the information necessary to formulate an effective production schedule at this level. In summary, activities at the operational level are the daily lifeblood of any organization and planners at this level must organize the daily work using detailed, short run planning horizons.

1.2 Rationale And Implications For Hierarchical Planning Systems

1.2.1 Rationale And Initial Implications

The review in section 1.1 of a hierarchical planning framework with three broad planning categories both invites and answers the question of why hierarchical planning approaches first developed. Each of the three planning levels address a complex set of linked, but different decisions. These decisions differ across many dimensions including length of planning horizon, level of detail required of planning decision support data, risks and costs of the decisions, and long term impact of the decisions to name only a few. Now let's evaluate the implications of the markedly different characteristics of the decisions required at each level from several different perspectives. First, consider the issue of attempting to build decision support planning models to facilitate efficient and effective decision making for all levels of production planning ranging from the strategic to the operational. It would essentially be impossible to construct "one" model to provide decision support for all planning decisions ranging from the strategic to the operational. (In fact, the hierarchical production planning literature abounds with discussions of the impracticality of attempting to develop single, monolithic models to address operational, tactical and strategic planning decisions [see e.g., Meal, 1984; Bitran and Hax, 1977; Hax and Meal, 1975; Mohanty and Krishnaswamy, 1984; and Mohanty and Kulkarni, 1987].) For example, such a model would have to provide outputs ranging from detailed short run production schedules to long run, multi-year projections of capacity requirements. Even if one could theoretically design a monolithic model to simultaneously address strategic, tactical and operational planning decisions, the issue of the "solvability" of such a model quickly arises. Operations research and management science techniques such as mathematical optimization and simulation play an increasingly important role in all levels and areas of production planning. Although extremely powerful and versatile, these techniques, as do most algorithms and "systems", ultimately have limitations beyond which they cannot effectively function. That is to say, it is still possible to design on paper an extremely complex mathematical model or system which cannot be solved by known mathematical algorithms. In the field of optimization, such models are often described as "np hard" type problems. Similar practical limitations or issues exist in the design of other planning systems besides optimization based systems. Thus, one implication of the multiple dimensions of production planning is that from a practical sense, it is very difficult to design one decision support model to address the complex hierarchy of decisions which must be made.

From a managerial perspective, the broad dimensions of production planning decisions virtually necessitate that multiple layers of management participate in the decision-making process. We observed in the previous section that senior management typically dominates the decision-making process at the strategic level, while less senior managers and supervisors generally assume the lead in tactical and operational decisions. Given this hierarchy of required management participation, it seems logical that a parallel hierarchy of decision support tools and approaches would exist to facilitate the management process. Thus, when viewed from both a management perspective and a decision support or systems perspective, one can argue that the breadth of activities associated with production planning demands a hierarchical approach.

Returning to the varying degrees of uncertainty and detail found in different decisions illustrates yet another key motive for developing a hierarchical planning approach. As observed in section 1.1, strategic decisions require a very aggregated level of information and they can influence a firm for many years into the future. In fact, the gestation period before the full impact of a strategic decision comes to bear can span several years (e.g., the decision to build a new plant). On the other hand, operational decisions require very detailed data and can impact a firm within hours or days (e.g., the decision of what products to produce on a production line next week). Obviously therefore, managers must make different types of decisions at different points in time. It would not make sense to decide whether to produce a blue or a red widget on a production line in a particular week three years from today. It is far too uncertain today to predict what color widget will be required on a particular week in three years. Further, the lead time required to implement this decision is probably only a few hours or days, and thus the firm can postpone this decision until the last minute. In contrast, it most certainly makes sense to decide today what type of plant with what production capabilities should be built, if a firm has concluded that it will need additional capacity within three years. Even though this decision may hold significant risk, the lead time of plant site selection and construction will dictate that the firm act far in advance. The preceding rather obvious example further illustrates that a firm or organization can best address the varying levels of risk, uncertainty, lead time and detail found and required in different decisions, by structuring all decisions in a hierarchical planning approach.

1.2.2 Implications For Design Of Hierarchical Production Planning Systems

The hierarchical framework reviewed in section 1.1 also suggests a number of implications about the design of a hierarchical planning system. Clearly, decisions made at each level affect the other levels of the hierarchy. Decisions made at the strategic level place constraints on the tactical level and ultimately the operational level, while decisions made at the tactical level constrain the operational level. At the same time, decisions made at the operational and tactical levels provide feedback to evaluate decisions made at the tactical and strategic levels, respectively. Decisions made at a lower level can also in certain cases indirectly force the direction of later decisions at a higher level. For example, consistently poor decisions in operational execution can result in inadequate customer service levels which ultimately may dictate future decisions at the tactical and/or strategic level. Thus, the fact that decisions at each level affect other levels requires that a firm design its hierarchical planning system with the proper blend of independence of decision-making at each level (*i.e.*, autonomy), yet at the same time with strong links between each level. The need for well-defined linkages between planning levels demands an integrated design approach. In fact, as discussed later in this book, a litmus test for evaluating a hierarchical planning system consists of determining whether a particular system has well established feedback and linkage mechanisms built into a closed loop, top-down and bottom-up system.

A second implication of the hierarchical framework discussed in section 1.1 concerns the need for a decomposition approach in hierarchical planning systems. Consider again the many facets of production planning from the strategic to the tactical to the operational. When simply judged as a planning problem, it becomes clear that one must decompose this problem into a number of subproblems in order to "solve it" (*i.e.*, create production plans for all levels). Algorithmically, this suggests that a hierarchical production planning system must consist of a number of linked models (or decision support systems) which provide inputs and/or outputs to each other. Further, one must integrate these decomposed sub-problems in such a fashion as to avoid excessive suboptimization. Again, as we will discuss later, another litmus test of the design of a hierarchical planning system is how effectively the design minimizes the potential level of suboptimization that a particular hierarchical system can produce.

In designing a hierarchical planning system, there are a number of critical questions and factors to consider. Figure 1.4 offers a brief review of selected key questions to consider in designing a hierarchical production planning system.

- How many planning levels should the hierarchical system have?
- What activities and processes should occur at each level of the planning hierarchy?
- What are the key problems which must be solved at each level?
- What planning models or decision support systems should exist to solve the key problems at each level?
- What methodologies or algorithms should be used to solve each problem?
- How should data and other informational inputs and outputs be aggregated and disaggregated at each hierarchical level?
- What are the key linkages between the different planning levels? How do decisions made at higher levels flow down to lower levels and how do decisions made at lower levels provide feedback to higher levels?
- How can the system be designed to best match the organizational structure and hierarchy of the firm?
- How can the potential for suboptimization introduced by a hierarchical approach be measured?
- What is the proper balance between designing a simple and manageable system, yet imbedding sufficient complexity and detail to realistically model the operation?
- How can the effectiveness of the overall system be measured?

Figure 1.4 Key Questions To Consider In Designing A Hierarchical Production Planning System

In considering how to address the questions posed in Figure 1.4, there exist a number of key characteristics of an operation which provide direction and guidance. For example, the existing management structure warrants strong consideration in the design of a hierarchical planning system. Clearly, the system must accomodate the hierarchy of the firm's decision-making process. Also, in any organization, regardless of whether or not it has a formal 'linked" hierarchical planning process, at the minimum there always exists some rough sequence in which planning decisions occur. (This process may not be consistent or properly integrated, however, it generally exists in some form.) A review of the existing sequential process - particularly the pros and cons of the existing process - represents a good starting point in the design of a true hierarchical planning system. Similarly, it is important the review the planning and decision-making structure of an organization from the perspective of how centralized or decentralized the current or planned future process is or will be. In firms of virtually any size, even relatively small operations, there exists a need to perform some planning and scheduling activities at the local level. On the other hand, it is also usually impossible to decentralize all planning and decision-making activities to the local levels. This in fact represents another key raison d'etre to develop a hierarchical planning approach. In terms of the design of a system, however, the level of decentralization in a firm's decision-making process can significantly influence the optimal design of a hierarchical planning approach (*i.e.*, the approach must accommodate the level of decentralization.)

The product structure of a firm represents one of the most commonly used inputs in the design of a hierarchical production planning system. A review of the product structure provides insights as to how one should aggregate individual end items into product families and product families into even more consolidated groupings. The hierarchical production planning literature abounds with alternative hierarchical product structures designed for specific problems and environments, and we will discuss some of these later in the book. The three level design of Hax and Meal (1975) consists of (1) end items, (2) product families, and (3) product types, and this probably represents the most often cited structure. However, every problem and firm has unique products with their own characteristics; and therefore, this demands a "blank sheet of paper" approach to the construction of the product hierarchy. In designing the product hierarchy, one must naturally consider the physical characteristics of the production process. This often drives the hierarchical design. Additionally, however, it is critical to determine what structure will facilitate the aggregation and disaggregation of data from and to the end item level without masking important information about the end item level. We will return to this subject later in the book.

The production technology and production process of a firm represents another key element to consider in designing a hierarchical production planning system. A hierarchical planning process which works well in a batch or continuous processing production environment may have little relevance to a job shop or fabrication type firm. Additionally, the lead times associated with the key production operations of a firm require careful consideration in the design of a hierarchical production planning system. Lead times of activities provide great insight as to the proper partitioning of planning problems into different levels of the hierarchy. (One way of viewing a hierarchical system is that long lead time decisions place constraints on shorter lead time decisions which in turn provide informational feedback to long lead time planning activities.)

The actual subproblems which require solution in a hierarchical planning system represent another important design consideration. Clearly, a firm must construct a system in which it can solve all of the subproblems or models that exist at each level of the hierarchy. Advances in operations research, management science and enterprise-wide planning software among others have increased the scope, sophistication and the complexity possible to address in each subproblem. While one should naturally take advantage of these advances, ideally, one should still design subproblems that at least remain conceptually relatively simple and solve in relatively short timeframes. As a firm progresses through the hierarchical planning design process it will continually wish to evaluate and re-evaluate the solvability, understandability and complexity of the planning system's individual subproblems.

1.3 A Generic Framework For Hierarchical Production Planning

To establish a foundation for later discussions of hierarchical supply chain planning and management, we present a generic example of a framework for a hierarchical production planning process. This generic example will provide a point of reference for later, more detailed reviews of specific hierarchical planning techniques and systems.

Figure 1.5 offers a general framework for hierarchical production planning which spans the strategic, tactical and operational planning and scheduling levels. Beginning at the strategic level, a firm must address such key issues as overall corporate objectives, market share and profitability goals, the desired mix of businesses and products, and so on. Planning decisions on overall corporate objectives then drive strategic manufacturing decisions. For example, market share and business or product mix objectives will strongly influence manufacturing capacity strategies.

At the strategic manufacturing planning level, the firm must address such issues as total planned production capacity levels for the next two, three or more years; the number of facilities it plans to operate, their locations, the resources the firm will assign to its manufacturing operations, and numerous other key long term decisions such as those discussed in section 1.1. Decisions made at the strategic production planning level place constraints on the tactical planning level. At the tactical level, typical planning activities include the allocation of capacity and resources to product lines for the next twelve months, aggregate planning of workforce levels, the development or fine-tuning of distribution and customer shipment sourcing plans, and other activities previously noted in section 1.1. Within the constraints of the firm's manufacturing and distribution infrastructure (-an infrastructure determined by previous strategic decisions), managers make tactical (annual) planning decisions designed to optimize the use of the existing infrastructure. Planning decisions carried out at the tactical level impose constraints upon operational planning and scheduling decisions. At this level, activities such as distribution resource planning (DRP), rough cut capacity planning, master production scheduling, shop floor control scheduling and the other decisions noted in section 1.1 occur.

The feedback loops from the operational level to the tactical level and from the tactical level to the strategic level represent one of the most important characteristics of a hierarchical production planning system illustrated in Figure 1.5. A true HPP system is a closed loop system which employs a "top down" planning approach complemented by "bottom up" feedback loops. Given the emphasis of HPP systems on evaluating capacity levels and imposing and/or communicating capacity constraints from higher levels down to lower levels, it is imperative that strong feedback loops exist. As is well known and we will explore in detail later, production plans which appear feasible at an aggregate level can often contain hidden infeasibilities that only manifest themselves at lower, more disaggregated levels. Without proper feedback loops imbedded into a hierarchical planning system, the danger that a firm will attempt to move forward with infeasible plans always exists. These infeasibilities often do not surface until a firm is in the midst of executing its operational plans and schedules.

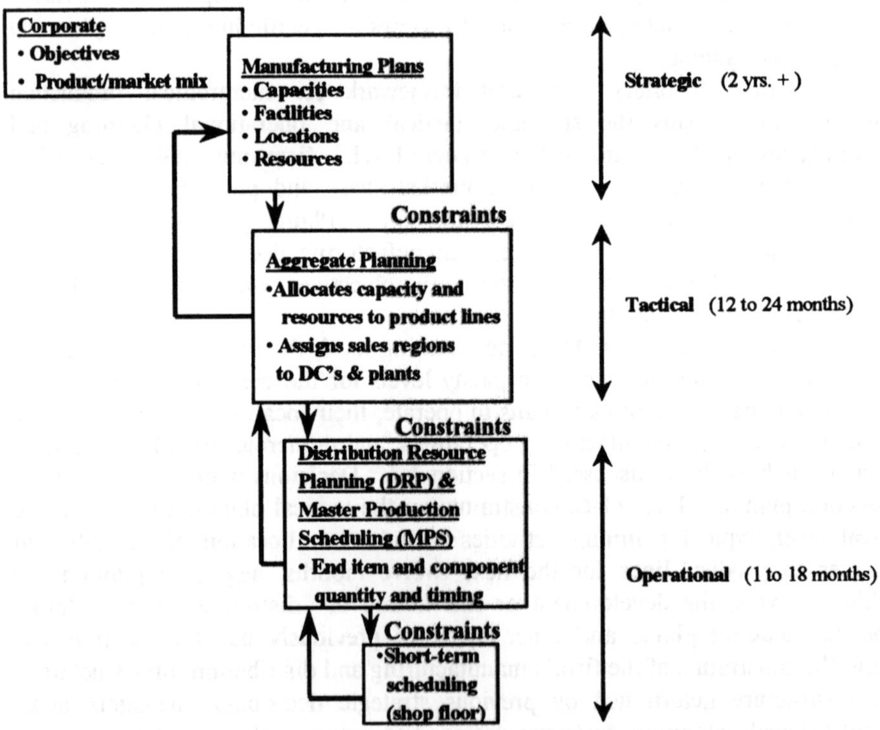

Figure 1.5 Hierarchical Production Planning Framework

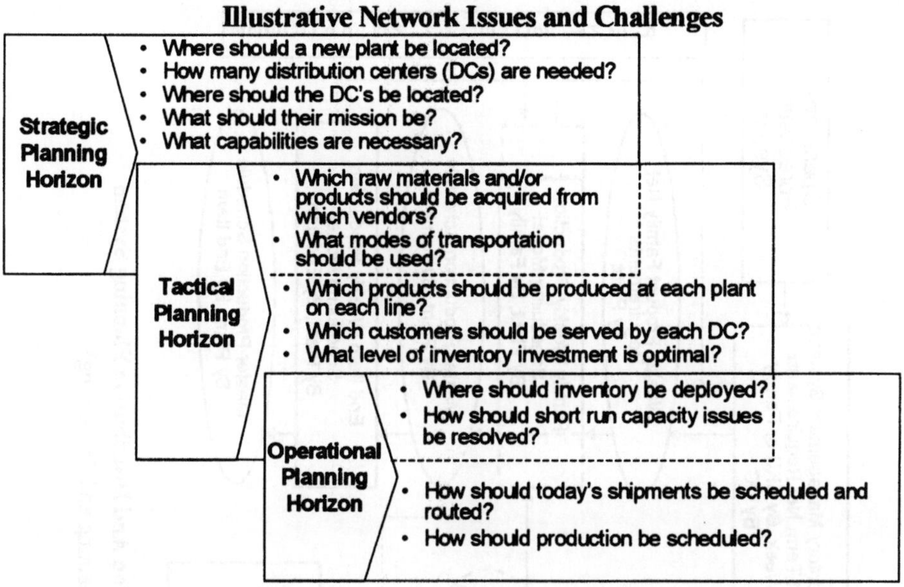

Figure 1.6: How Network Decisions Fit Into A Planning Hierarchy

The hierarchical framework presented in Figure 1.5 is generic in the sense that although individual HPP systems may differ dramatically, most systems are designed within this or a similar general framework. Figure 1.6 recaps some selected generic decisions previously described that a HPP system constructed within this framework will address and it displays how these decisions fit into a planning hierarchy.

With a general HPP framework in place, the next step in constructing a HPP system consists of designing its actual components. At this point it becomes more difficult to describe a "generic" HPP system. This results from the fact that actual implementations of HPP systems all possess unique characteristics reflective of the particular firm, industry and/or environment where the HPP approach is applied. Further, it is not uncommon to observe HPP systems which focus primarily upon only the tactical and operational planning and scheduling levels (see e.g., Dempster et. al., 1983, and Liberatore and Miller, 1985). Nevertheless, for purposes of establishing a point of reference regarding HPP implementations, we conclude this section with a schematic for a generic HPP implementation, focusing on the tactical and operational levels.

Figure 1.7 illustrates a generic tactical and operational HPP system. While admittedly not all HPP implementations follow this design, nor contain these specific components, most HPP's perform the majority of the functions depicted in this diagram. Because twelve to eighteen month rolling planning horizons represent the most common tactical planning horizons, we illustrate an integrated annual/tactical and operational planning and scheduling system. Note that the boxes with broken lines in Figure 1.7 indicate elements of the annual planning

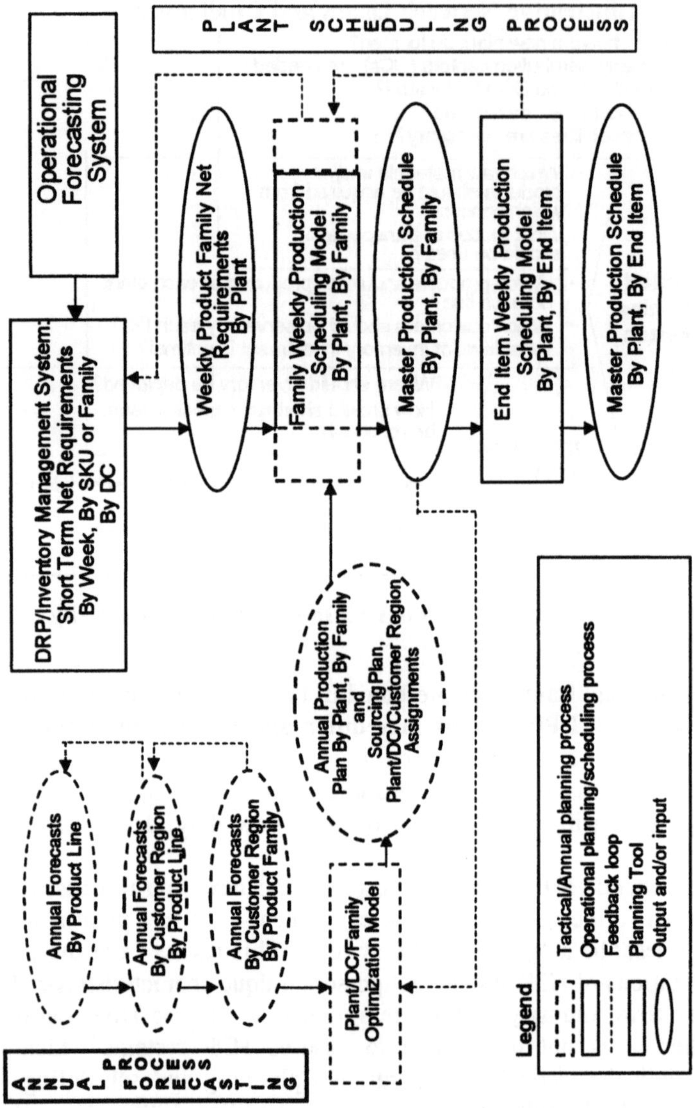

Figure 1.7 Integrated Production And Distribution Planning System

(Tactical/Annual Planning and Scheduling)

process, while the boxes with solid lines constitute the operational planning and scheduling process. Beginning with the annual planning process, let's review this integrated tactical and operational HPP system.

The annual process commences with the development of annual or twelve month rolling forecasts by major product line.[1] Techniques used to generate forecasts at this level vary widely and include econometric forecasting, exponential smoothing, time series based forecasting and others. Importantly, exogenous market intelligence and sales judgement factors often provide critical inputs at this level. Thus, a firm's ability to structure a process which effectively integrates quantitative techniques and qualitative inputs represents a key determinant of the accuracy of major product line forecasts. The next step consists of disaggregating total product line forecasts into product line forecasts by customer region. Methodologies for generating these lower level product line forecasts again vary widely and range from simple techniques such as using recent historical ratios of each customer region's demand to total demand, to more complex quantitative approaches (see Oliff, 1987 and Miller, 1986). The size of individual customer regions strongly influences the approach at this level. For example, in a global HPP system, customer regions may represent geographic regions such as countries or a region of countries. Global applications will more likely employ more sophisticated methods to develop product line forecasts. In fact, many firms choose to develop global product line forecasts from aggregations of individual country product line forecasts. (Note the feedback loop between the product line forecast by customer region and the total product line forecast.) On the other hand, in the case of an HPP system covering a smaller geographic span (e.g., a domestic U.S. system) where customer regions represent relatively small entities (e.g., parts of individual states), it is more common to observe simpler methods employed to disaggregate product line forecasts into product line forecasts by customer regions (see e.g., Miller, 1986).

The next step in the annual planning process entails breaking customer region product line forecasts into product family forecasts by customer region. Again, there exist a range of methods to derive such forecasts and we will review several of the principal techniques in subsequent chapters. Further, the sequence or hierarchy by which firms obtain product family forecasts by customer region can also vary. For example, some HPP applications employ a product line to product family to product family by customer region disaggregation sequence rather than the sequence depicted in Figure 1.7. There is no universally optimal approach for all forecasting applications and often several schemes will work well. Regardless of the disaggregation sequence, however, ultimately the methodology shown in Figure 1.7 requires the generation of product family forecasts by customer region.

The next step in the annual planning process, namely running a plant/distribution center/product family assignment model (*i.e.*, optimization

---

[1] For this example, we assume a product hierarchy consisting from top to bottom of (1) major product line, (2) product family, and (3) end item where a product family represents an aggregation of similar end items, and a product line represents an aggregation of similar product families. We will return to the topic of product hierarchies again in later chapters.

model) drives the requirement for product family forecasts by customer region. Linear and/or mixed integer programming optimization models represent a standard decision support tool utilized at the tactical/annual planning level of a HPP system. Annual optimization models most frequently define products at the product family level (see e.g., Hax and Meal, 1975). Further, assuming that the HPP application involves multiple manufacturing or distribution locations, these annual optimization models define demand by appropriate geographic regions. Hence, the need for annual product family forecasts by customer region. If the annual model contains multiple time periods (e.g., months or quarters), then it naturally requires the annual product family forecasts by customer region, by time period (*i.e.*, one additional level of disaggregation). Figure 1.8 highlights some of the key outputs derived from the annual plant/DC/family optimization model.

- **Distribution plan**
  - Assigns sales regions to DC's and plants (*i.e.*, a sourcing plan)
  - Establishes plant to DC and inter-DC shipping (supply) plans
- **Annual production plan, for each plant, by time period**
  - Establishes planned operating rates for each plant (*i.e.*, capacity utilization rates)
  - Assigns production product mixes for each plant (by defined product families)
- **Inventory plan for planning horizon, by time period**

Figure 1.8 Outputs From Annual Plant/DC/Family Optimization Model

As Figure 1.8 indicates, the annual optimization model creates a production plan by plant, by family, by time period and sourcing assignments by plant, by distribution center, by customer region.

As shown in Figure 1.7, the last major step of the annual planning process consists of evaluating the annual optimization model's production plans for each plant in a plant level weekly product family production scheduling model. Briefly, this model helps to determine whether the production plans developed for a plant by the annual model remain feasible when evaluated at a greater level of detail than possible in an annual model. The weekly production scheduling model produces a master production schedule for a plant at the family level. It is not unusual to find that production plans which appeared feasible at the annual level manifest infeasibilities when evaluated at the weekly level. This results because annual integrated manufacturing/distribution models typically do not include changeover costs, nor do they define time buckets at weekly levels. A combination of factors explain why annual models usually do not contain this level of detail including:

1. it generally does not make sense to define sourcing plans at a weekly level (e.g., one usually does not want to have a customer region's source DC or plant change from week to week), and

2. mathematical complexity often precludes attempting to solve such a model, even in those occasional cases where it makes sense to include great detail in an annual planning model.

Thus, in Figure 1.7, the feedback loop shown between the output of the weekly plant product family scheduling model and the annual optimization model represents a critical part of the annual planning process. Clearly, a firm must develop an annual production plan which each of its plants can feasibly implement at the operational level. The feedback loop from the weekly model to the annual model assures this.

This completes the integrated annual hierarchical generic planning process. Next we briefly review the operational planning and scheduling process. Based upon the tactical (annual) planning process, sourcing patterns have been defined (e.g., what plant serves which DC and what DC serves which customer region). The operational planning and scheduling system executes these decisions on a day-to-day and week-to-week basis.

Figure 1.7 illustrates a standard operational planning and scheduling system. A forecasting system, frequently time-series, exponential smoothing based, generates forecasts at the item or sku level. These forecasts feed an inventory management system, often a distribution resource planning (DRP) system, which generates net requirements by week, by sku and/or by family for each distribution center. The DC's place their weekly net requirements upon their source plants and this creates the weekly net requirements by end item, by family for a manufacturing plant. These weekly net requirements drive the family weekly production scheduling model. (Note that this short run focused scheduling model can be defined in different time buckets [e.g., days or two-week periods] if appropriate.) Figure 1.7 shows a feedback loop flowing from the weekly scheduling model back to the inventory management module. Should the weekly scheduling model determine that a plant cannot meet the weekly family net requirements demanded of it, this feedback loop communicates the infeasibility issue. An iterative process would then ensue whereby either the original net requirements are modified until a feasible plan can be produced, or excess production requirements are addressed by such options as overtime production or the offloading of some net requirements to other plants.

The final major step outlined in the generic process of Figure 1.7 is the development of a plant's master production schedule by end item. The plant's weekly master production schedule by family provides the input which drives the end item scheduling model. This model produces a traditional master production schedule of end items by weekly time bucket. Note that a feedback loop also flows from the master production end item schedule back to the weekly family production scheduling model. This feedback loop is required because it is possible that a schedule which appears feasible at the weekly, family level may mask infeasibilities that become evident at the item level. Chapter 6 will offer an example of how this can occur.

In summary, this section has presented a generic hierarchical tactical and operational production planning system. As noted from the outset, not all HPP implementations exhibit this exact framework or set of components. Nevertheless, HPP implementations have similar general frameworks to that presented here, and contain many of the individual components illustrated in this section. This generic framework will provide a general reference point as we move on to examine the numerous aspects of hierarchical production and supply chain planning and scheduling.

1.4 Objectives Of The Book

The remainder of this book builds towards our ultimate objective of developing a comprehensive review and understanding of how hierarchical planning techniques and principles can contribute to the management and planning of supply chain activities. The pursuit of this objective will require us to review current and past hierarchical production planning techniques and implementations. We will also study hierarchical planning approaches used in different areas of supply chain management (*i.e.*, approaches employed for activities other than production planning). This broad based review will provide insights on methodologies and hierarchical planning techniques employed in certain aspects of supply chain management that have applicability in other areas of the supply chain. Thus, an additional objective of this book is to stimulate a cross-pollination of ideas and approaches through this study.

Another objective of this book is simply to highlight the importance of employing a well-constructed hierarchical approach towards supply chain management regardless of whether one is addressing classical hierarchical problems (*i.e.*, production and distribution planning) or other problems that do not normally fall within the purview of HPP. This objective holds particular importance because of the recent advances in commercially available, "user-friendly packaged" software systems which facilitate hierarchical production and supply chain planning and scheduling. These systems are sometimes described as "enterprise-wide" planning software in the current venacular. For industry practitioners attempting to employ this software effectively, it is critical to understand the philosophy and classical techniques of hierarchical production and distribution planning. This knowledge allows the practitioner to implement this relatively new genre of "user-friendly" software in the manner which best meets the practitioner's planning needs. Without a strong grounding in hierarchical planning methodologies, the practitioner may simply implement the commercial software in its "vanilla" or standard configuration regardless of whether this configuration best addresses the practitioner's particular problem. (*I.e.*, The practitioner may fall into the trap of blindly following the software manufacturer's generic design.)

Finally, the field of hierarchical production planning is one that has flourished for decades in the confines of academia and the research world. It is hoped that this book will make a small contribution towards providing a wider base of appeal for, and for facilitating more actual implementations of hierarchical planning approaches across all supply chain management activities.

1.5 Organization Of The Book

The outline for the remainder of this book is as follows. Chapter 2 begins with an introduction to some of the classical approaches used in hierarchical production and distribution planning. In this chapter, we will present several case studies of implementations of integrated, hierarchical manufacturing and distribution planning systems including those of Hax and Meal (1975) for a tire manufacturer, Glover et. al. (1979) for a large chemical fertilizer company, and Liberatore and Miller (1985) for a ceramic tile manufacturer. This chapter, which focuses on how to manage broad-based implementations that integrate production and distribution planning, includes recommendations of techniques used for communicating key planning information across the organization. Chapter 3 takes a narrower view than Chapter 2, focusing strictly on production planning and scheduling systems. In this chapter, we study both annual production planning methods and short run plant scheduling techniques, and we evaluate hierarchical approaches for linking long run and short run planning activities by means of feedback loops and algorithms. This chapter also evaluates how material resource planning (*i.e.*, MRP) fits into the hierarchical production planning paradigm.

Chapter 4 examines the activities of distribution and transportation planning and scheduling. Topics discussed in this chapter include warehouse location decision-making, multi-echelon multi-mode integrated distribution and transportation planning, strategic rail network planning, and mode and carrier choice selection. In studying transportation, we present alternative methodologies to facilitate mode choice selection and we examine such issues as how to aggregate and disaggregate transport lanes and products for annual transport negotiating and planning. This chapter also discusses short run transportation scheduling methods and how to link short run transportation scheduling and annual transportation planning.

In Chapter 5, we explore the use of hierarchical planning techniques in the design and management of warehouses. Topics reviewed in this chapter range from strategic and tactical decision support planning for warehouse facility design and technology selection to operational decision support methods to allocate items to aisles and zones. Again, we focus on the benefits of employing an integrated approach for all aspects of warehousing ranging from strategic facility and technology design to day-to-day operational execution.

Chapter 6 explores inventory management and forecasting, two linked activities of supply chain management which commonly utilize hierarchical decision support planning tools. In this chapter, we consider such classical issues as product hierarchy decisions, the stratification of inventory classifications for planning and management purposes, and how to use integrated top-down and bottom-up forecasting methodologies to obtain organizational alignment and more accurate forecasts. Chapter 6 also presents a hierarchical decision support methodology for evaluating a firm's inventory position over multiple distribution echelons. With the conclusion of Chapter 6, we complete our review of hierarchical decision support planning for the core components of the supply chain.

In Chapter 7, we consider several hierarchical performance measurement frameworks for monitoring and evaluating supply chain activities. The chapter starts with a review of performance measurement systems used to evaluate both a firm's supply chain performance and its overall performance. This chapter illustrates how one can integrate such popular firm performance measurement systems as the balanced scorecard (Kaplan and Norton, 1993 and 1996) into a hierarchical performance measurement system. Further, we offer a methodology which uses both the balanced scorecard and an operations research based methodology, the analytical hierarchy process (Saaty, 1996); to develop integrated performance monitoring indices that measure a broad spectrum of activities ranging from overall firm performance to specific supply chain activities such as distribution and manufacturing.

Finally, in Chapter 8, we summarize the wide-spread applications of hierarchical planning techniques found across all areas of the supply chain. Based upon the perspective provided by Chapters 2 through 7, Chapter 8 concludes with an assessment of the benefits and rationale of this methodology.

Chapter 2

Managing Operations On A Network: Integrated Manufacturing And Distribution Planning

2.1 Introduction

Hierarchical production planning encompasses an extremely broad spectrum of planning problems ranging from hierarchical approaches for scheduling a single plant or perhaps even a single production line to planning and scheduling operations over an entire global supply chain. In this chapter, we focus on hierarchical planning implementations which provide guidance for managing and scheduling integrated manufacturing and distribution operations over a multi-echelon network. Specifically, we will review several case studies of hierarchical production and distribution planning implementations reported in the literature over the last several decades. These case studies span several different industries and provide insights into the common elements found in most successful HPP implementations. At the same time, these case studies will also serve to illustrate the significant differences typically found from one HPP implementation to the next. We will begin this chapter with a review of perhaps the seminal HPP implementation; namely, the HPP system developed by Hax and Meal (1975) to assist in managing the manufacturing network of a large tire producer in the United States. Other HPP case studies reviewed include one developed for a large chemical fertilizer manufacturer (Glover et. al, 1979) and one established for a ceramic tile manufacturer (Liberatore and Miller, 1985). All of these cases share the common characteristic that they represent HPP implementations designed to plan and schedule manufacturing and distribution operations in an integrated fashion over a large scale network.

To successfully implement an HPP system over a large manufacturing and distribution network requires the development of an effective strategy to communicate planning inputs, outputs and decisions to all concerned parties. A well designed HPP system that lacks a complementary communications forum will likely not deliver the full benefits that the HPP could otherwise offer. Thus, after reviewing case studies, this chapter concludes with a review of an approach for disseminating key information about the HPP planning system throughout an organization.

2.2 Case Study: An HPP Implemented By A Tire Manufacturer[1]

In the mid-1970's, Arnoldo Hax and Harlan Meal designed and implemented a hierarchical production planning system for a large U.S. tire manufacturer. At the time, the firm had four plants located in the eastern, north central, south central and western regions of the U.S. respectively. Each plant manufactured all the high volume tires of the firm's product line, however, only one or two plants produced the slower moving items. Each plant had a distribution center that was associated with that plant. The distribution centers received products both from their respective "associated" plant as well from the other plants in the network. Thus, each distribution center shipped mixed, full product line orders to its respective customers. This firm, similar to other tire manufacturers, experienced a significant peaking of demand for its products during a year. Further, the peak period demand rates typically exceeded the firm's period manufacturing capacity. Thus, to operate efficiently, its planning systems had to plan production and distribution network-wide in a coordinated, forward-looking approach.

2.2.1 Case Problem, Objective And Additional Background

The firm's management believed that a significant opportunity existed to reduce inventory and costs while at the same time improving customer service (*i.e.*, order fill rates). Hax and Meal (1975) described that:

> The manufacturing firm .... was faced with a common set of symptoms resulting from the use of poor and uncoordinated rules in the management of production planning and scheduling decisions. Inventories were both too high for certain items and too low for other items. Thus, inventory investments were large but poor customer service was provided in many standard stock items. Moreover, high production costs resulted from frequent interruptions in the manufacturing process as a consequence of run-outs. This, in turn, led to reduced production capacity and missed promise dates on special customer orders.

In order to improve inventory utilization, customer service and manufacturing efficiency, the firm decided to implement a network-wide, hierarchical production and distribution planning system to coordinate activities among its four plants. We will observe shortly that this system assured an appropriate balance between centralized and decentralized planning and scheduling activities. Prior to this HPP implementation, each plant had operated very autonomously. Meal (1984) noted that:

---

[1] This section draws heavily upon the reports of this case study found in Hax and Meal (1975), and Meal (1984). The description of the firm's background and implementation approach is based on the aforementioned articles.

> ...planning was decentralized... After senior management issued an initial directive indicating which items would be produced at each location, plant managers were left to plan production of the stock needed to fill the orders received from customers and other plants....Plant managers usually....(built) stock of high-volume items they were confident of selling. Thus, many small runs of slow-moving items were left for the peak season, when capacity was at a premium. If the needed tires were not produced locally, each distribution center ordered them from the other company plants, placing these orders as late as possible to keep local inventory as low as possible. This practice led to a high volume of small orders during the peak season, thus aggravating an already difficult situation.

Thus, the mission and objective of the new HPP system was to improve the firm's operating practices by coordinating production planning across all plants in an integrated approach that allowed decisions to be made by the appropriate personnel at the appropriate location.

2.2.2 Overview Of Hierarchical Planning System

The HPP implemented by Hax and Meal decomposed the tire manufacturer's production planning and scheduling process into four major "linked" decision levels. We will first review this system at a summary level, and then consider each system component in additional detail. Table 2.1 summarizes the four level design.

The first major decision consisted of determining the product mix assignments of each of the four plants for the next year or so (*i.e.*, determining which plants would produce what products in what quantities). A mixed integer program given the name the plant/family assignment model was developed to provide analytical support for this decision. The analysis at this level also evaluated long-term production capacity requirements and projected utilization rates. Specifically, in addition to product mix assignments, the plant/family assignment model would also determine if incremental capital investments were required to build additional capacity at plants. Thus the first level of the HPP evaluated the entire tire manufacturing network simultaneously to develop plant product mix assignments, plant capacity utilization plans and plant capital investment plans for the next twelve months. Note that this infrastructure utilization planning took place once a year. With high-level infrastructure type decisions made, the next set of major decisions in the HPP could take place.

Table 2.1 Hierarchical Planning Process Overview

System Name	Decision	Decision Method	Decision Level	Frequency Of Updates To Decisions
Plant/Family Assignment Subsystem	• Products are assigned to plants • Incremental capital investments necessary to support production capacity requirements are determined	Mixed-integer programming	Corporate	Annual
Seasonal Planning Subsystem	Monthly production plans and seasonal inventory accumulation plans are developed for each product type	Linear programming	Plant	Monthly
Family Scheduling Subsystem	Detailed production schedules for each product family are developed	Standard lot size and run length methods	Plant	Monthly
Item Scheduling Subsystem	Detailed production run quantities for each item in each family are developed	Standard lot size and run length methods	Plant	Monthly

At the second level of decision-making, the tire manufacturer generated monthly production plans and seasonal inventory accumulation plans by product type for each plant. (As we will discuss shortly, product types represented the highest level of aggregation of products in the Hax-Meal product hierarchy.) A linear programming model titled the seasonal planning subsystem or model assisted this process. This planning activity took place at the beginning of each month for each plant as the manufacturer would update previous plans based upon the latest inventory and demand data. Importantly, note that this second level of decision-making (*i.e.*, the high level scheduling of a plant by product type) took place at the plant level. This contrasts with the first level where the decision-making was based upon a network-wide, corporate perspective.

The third component of the HPP was called the family scheduling subsystem, and at this level, detailed production schedules for each product family within each product type were developed. Using standard lot-sizing and run length inventory control methods, planners would develop run schedules for each product family at each plant. This process also occurred monthly and would take place once planners had completed the level two monthly decision-making process.

The fourth and final major HPP component, the item scheduling subsystem, generated a detailed production run schedule at a plant for each item. On a monthly basis, again using standard lot-sizing and run length algorithms, the item scheduling subsystem would take the decisions made by the family scheduling subsystem as inputs, and then transform these product family production schedules into item production schedules. This activity would occur at each plant, and on the basis of the item schedules developed, each plant would run its operations.

This completes an initial overview of the HPP system developed by Hax and Meal for the tire manufacturer. To facilitate a more detailed examination, we first must discuss the product hierarchy developed for this planning system.

2.2.3 Product Structure Hierarchy

Hax and Meal created a three level product hierarchy to support their HPP system. Defining an appropriate hierarchical product structure represents an essential prerequisite to the successful implementation of any HPP system, and the Hax-Meal hierarchy has been used and cited for many years. Table 2.2 offers an overview of this product structure. Items represent the lowest level in the structure and they are the finished goods units which consumers purchase. The firm's total product line contained approximately 10,000 items in total.

Table 2.2 Product Structure Hierarchy

Product Level	Level Of Aggregation	Approximate Number	Common Characteristics Used To Facilitate Aggregation
Item	None	10,000	No Aggregation
Family	Is a group of individual items	200	• Items in a family share tooling and setup costs • Items in a family are usually produced together
Type	Is a group of families	5	• Families (and therefore items) in a type have the same seasonal demand pattern • Families in a type can all be produced at the same production rate

The product family is the second level in the three-part structure. In the Hax-Meal framework, families represent groups of items that share two common characteristics. First, all items in a family share the same tooling and setup costs. When a line is setup to produce a particular product family, by definition, it can produce any item in that family. It may be that minor setups or cleanups are necessary to switch from production of one item to another within a family. However, no family type changeovers are required. The second (and related) common characteristic of all items in a family is that they were usually all produced together. This results because family changeovers were usually more significant in terms of costs and time, and therefore, it was quite common (although not required) to produce all items in a family when a line was setup to manufacture a particular family. The tire manufacturer had approximately 200 product families in its product line.

Product type is the highest level of aggregation in the Hax-Meal framework. Product types are groups of product families that: (1) have similar seasonal demand patterns, and (2) can all be produced on the same production line or lines at the same production rate. Hax and Meal required only 5 product types to describe the manufacturer's entire product line. This proved of great benefit for modeling purposes because it greatly reduced the potential number of demand constraints that they had to define in their seasonal planning models. (In general, the process of product aggregation reduces the number of variables and constraints required in planning models, an important benefit of the aggregation.)

2.2.4 Hierarchical Planning And Scheduling System

Hax and Meal utilized a variety of planning and scheduling techniques in designing the tire manufacturer's hierarchical planning system. Figure 2.1 provides an overview of the entire system as originally reported by Hax and Meal (1975). As this figure illustrates, the system contained four major planning subsystems, each of which generated outputs that became inputs to the next subsystem in the hierarchy. This represents a standard characteristic of a hierarchical planning system. To understand how this system operated, we will now consider each of these major subcomponents in more detail.

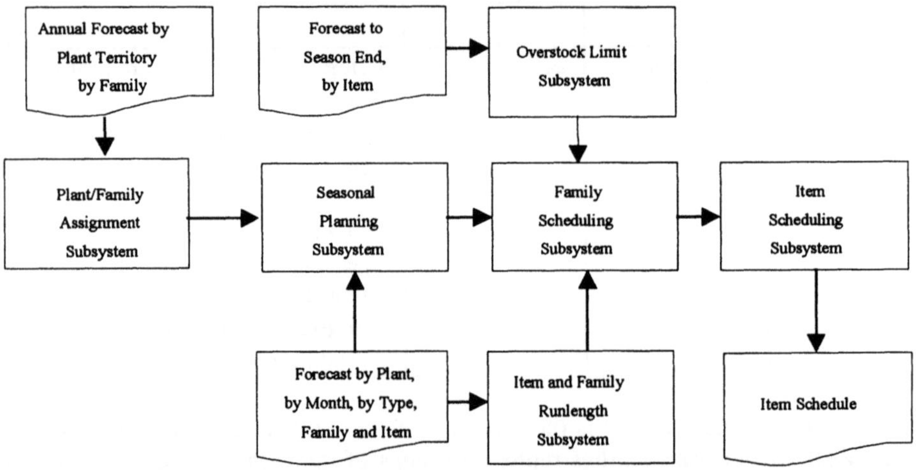

* Reproduced from Hax and Meal, 1975.

Figure 2.1 Overview Of Hierarchical Planning System

The Plant/Family Subsystem

As previously noted, the plant/family subsystem determines the product/production mix assignment for each plant and also it determines whether a plant will require additional capital investment in equipment to produce its assigned product mix. The core of this subsystem consists of a mixed integer programming model which minimizes incremental capital costs, variable manufacturing costs, and interplant transportation costs. Table 2.3 displays the notation used to formulate this model and equations (2.1) - (2.4) describe the model.

Table 2.3 Notation For The Plant/Family Assignment Model

$C_m =$	incremental capital investment required to produce the Family at plant location *m*. (This cost is independent of the other Families produced at the plant.)
$g_{mn} =$	cost of making one unit of the Family at plant *m* and transferring it to plant *n* for delivery to the final customer. If *m*=*n*, the transfer cost is zero.
$F_m =$	forecast demand in plant territory *m*, during the economic life of the investment (an average of three years).
$x_m = \{$	1, if the Family is assigned to plant *m*; 0, otherwise
$y_{mn} =$	production at plant *m* to fill demand in plant *n* territory.

Plant Family Assignment Model

Minimize: $$\sum_m C_m x_m + \sum_m \sum_n g_{mn} x_m y_{mn} \qquad (2.1)$$

Subject to: $$\sum_m x_m y_{mn} = F_n \qquad \text{for all } m, n \qquad (2.2)$$

$$y_{mn} \geq 0 \qquad \text{for all } m, n \qquad (2.3)$$

$$x_m = 0, 1 \qquad \text{for all } m \qquad (2.4)$$

The plant/family assignment model determined the products that a plant would produce by balancing the variable costs of production, interplant (*i.e.*, inter-distribution center) transportation, and interplant handling – all represented by (g_{mn}), and the incremental capital costs (C_m) required to set up a plant to produce a family. Thus, this model developed a network-wide integrated plan that evaluated and minimized both production and distribution costs. In order to develop this plan, the model required forecasts by product family, by plant territory. Further, because this model included capital investment costs on production equipment that had about a three year economic lifespan, Hax and Meal employed three year product family forecasts by plant territory. In contrast to an assignment model for a ceramic tile company that we will review in section 2.4, this model did not determine each plant (and distribution center's) territory at the same time as the product family/plant assignments were made. Rather, this model assumed that plant/DC territories were defined, and then minimized production and distribution costs, given the previously made plant/DC territory assignments.

There were several other notable aspects of this plant/family assignment model. First, the model did not contain production capacity constraints. After a model run or runs developed a proposed plan, the planners would evaluate whether the solution violated any capacity constraints. If the original solution contained infeasibilities, the planners would then have to reassign selected product families to develop a feasible plan. A set of penalty costs associated with reassignments served to guide the reassignment process. Hax and Meal explained that they did not consider "plant expansion" as an alternative to reassignments because the expansion costs far outweighed the reassignment costs. They also noted that although capacity constraints would have represented a desirable enhancement to this model, at the time of this system's inception (the mid-1970's), mixed integer programming had not advanced to the point where it could solve such a model. (This would not represent a limitation today.) A second important point to note is that the plant family assignment model evaluated a planning horizon of up to three years as a one period model. Hax and Meal also noted that ideally they would have formulated this model as a multi-period (*i.e.*, a dynamic) model so that it could address the tire manufacturer's highly seasonal demand. Again, mixed integer programming limitations at the time of this model's implementation precluded such an approach.

The Seasonal Planning Subsystem

The second major component in the planning hierarchy, the seasonal planning subsystem, determined the production and inventory plan for each plant, by defined time period, by product type. Recall that product types represented aggregations of product families that had: (1) similar seasonal demand patterns, and (2) similar production rates and costs (and therefore, similar inventory carrying costs). Hax and Meal developed a linear programming model to generate these seasonal product type plans. Table 2.4 and equations (2.5) – (2.11) display the notation and formulation of the seasonal planning model.

The seasonal planning model determined the production and inventory plan for each product type by minimizing (or balancing) overtime costs ($CO_{it}O_{it}$) and inventory holding costs ($CI_{it}I_{it}$). Hax and Meal note that in this case the tire manufacturer was essentially committed to its regular time labor over the planning horizon, and therefore, regular production hours (R_{it}) appear in this formulation as constraints only (2.6, 2.8, 2.9) without a cost coefficient. However, it is important to note that in aggregate workforce planning type models with planning horizons of twelve or more months, it is also quite common to observe regular time production included as a variable in the objective function with a cost coefficient.

Table 2.4 Notation For The Seasonal Planning Model

R_{it}	= hours of regular production of Type *i* to be scheduled during time period *t*
O_{it}	= hours of overtime production of Type *i* to be scheduled during time period *t*
R_t	= total hours of regular production available during time period *t*
O_t	= total hours of overtime production available during time period *t*
I_{it}	= inventory of Type *i* on hand at the end of time period *t* (units)
r_i	= production rate for Type *i* (units / hr)
D_{it}	= forecast demand of Type *i* during time period *t* (units)
CO_{it}	= cost of overtime production of Type *i* during time period *t* ($ / hr)
CI_{it}	= inventory holding cost for Type *i* during time period *t* ($ /unit per period)
SS_{it}	= safety stock required for Type *i* at end of time period *t* (units)

Seasonal Planning Model

Minimize: $$\sum_i \sum_t CO_{it} O_{it} + \sum_i \sum_t CI_{it} I_{it} \tag{2.5}$$

Subject to: $$\sum_i R_{it} \le R_t \quad \text{for all } t \tag{2.6}$$

$$\sum_i O_{it} \le O_t \quad \text{for all } t \tag{2.7}$$

$$r_i(R_{it} + O_{it}) - I_{it} + I_{i,t-1} = D_{it} \quad \text{for all } i, t \tag{2.8}$$

$$R_{it} \ge 0 \quad \text{for all } i, t \tag{2.9}$$

$$O_{it} \ge 0 \quad \text{for all } i, t \tag{2.10}$$

$$I_{it} \ge SS_{it} \quad \text{for all } i, t \tag{2.11}$$

The planning horizon of the seasonal planning model extended a rolling fifteen months into the future and contained seven subperiods: three individual months (periods 1, 2 and 3) and four quarters (periods 4, 5, 6 and 7). The firm employed a planning horizon of fifteen months to assure that it properly planned for seasonal inventory buildup requirements at all times. The tire manufacturer ran this model monthly with the latest updated forecasts and inventory positions, and adjusted its plans as appropriate based on the model's outputs. The minimum inventory targets for each product type for each period (2.11) represented the sum of the end of period safety stock requirements of each end item in each product type for each of the seven model periods.[2] Thus a solution to the seasonal planning model provided a plan of the number of production hours to be devoted to each

---

[2] To effectively utilize a model such as the seasonal planning model, it is important to decide how to define beginning inventories at the start of the planning horizon. Chapter 6 presents an algorithm and an approach to determine beginning of planning horizon inventory positions.

product type over the next fifteen months in order to satisfy demand and to meet minimum safety stock requirements.

The Family Scheduling Subsystem

With production hours for each product type determined for the planning horizon by the seasonal planning model, Hax and Meal required another subsystem to schedule the actual product families within each product type. The family scheduling subsystem played this role. The family scheduling subsystem did not have a core model, but rather consisted of a series of logical steps and formulas that produced a product family schedule. Hax and Meal described this system as follows:

> [the system] first determines which families in the type must be run in the scheduling interval in order to meet item service requirements; second, set initial family run quantities so as to minimize cycle inventory and changeover costs; third, adjusts the run quantities of the families scheduled to use all the production time available to the type, while fourth, observing item overstock limits.

To determine which families in a type to schedule, Hax and Meal first evaluated the "runout" time of each family in every type scheduled by the seasonal planning model. The runout time of a family was defined as the earliest point at which the inventory of any item in a family dropped to the minimum safety stock target for that item. Equation 2.12 calculates the runout time (RT_j) for each item k of each family *j*.

$$RT_j = \min_k\left[\left(A_{kj} - S_{kj}\right)/F_{kj}\right] \qquad (2.12)$$

where

A_{kj} = available stock of the k_{th} Item in Family *j*

S_{kj} = Item Safety Stock level required at the end of the scheduled period to meet the service requirement

F_{kj} = forecast usage rate of k_{th} Item

The family scheduling subsystem focuses on the first (*i.e.*, next) month in the planning horizon. Thus, equation (2.12) served to identify any item in any family that had insufficient beginning inventory to cover the next month of the planning horizon. Any family with an item that fell into this category had to be scheduled. The family scheduling subsystem next developed a preliminary run size quantity for each family to be scheduled. This initial quantity (RQ_j) represented the sum of the minimum quantity of each item in a family that must be produced as defined by equation (2.13).

$$RQ_j = \sum_k RQ_{kj} = \sum_k \min\left[Q_{kj}\left(L_{kj} - A_{kj}\right)\right] \tag{2.13}$$

where:

Q_{kj} is the minimum production lot size of an item k (previously determined based on standard item lot sizing principles), and L_{kj} is the overstock or maximum inventory limit for item k. (The maximum overstock limit for an item was determined analytically by evaluating such factors as the cost of holding inventory for an extra period versus the cost of an extra family changeover.)

The remainder of the family subsystem scheduling logic consisted of adjusting, if necessary, the preliminary family run sizes to their final actual run lengths. Briefly, the system evaluated whether the sum of the production requirements of the preliminary family run quantities of each family in a type exceeded the production capacity previously allocated to each family type by the seasonal planning subsystem. If no capacity violations existed for a product type (*i.e.*, $\Sigma_j\ RQ_j < r_i\ [R_i + O_i]$) then equation (2.14) determined the final family run quantities for that type (RQ_j*).

$$RQ_j^* = \min\left\{\sum_k\left(L_{kj} - A_{kj}\right), RQ_j + \left[r_i\left(R_i + O_i\right) - \sum_j RQ_j\right]\sum_k\left(L_{kj} - A_{kj}\right) / \sum_j\sum_k\left(L_{kj} - A_{kj}\right)\right\} \tag{2.14}$$

The second term in (2.14) simply served to assure that if the sum of the preliminary run quantities of all the families in a type did not exhaust the capacity previously allocated to that type, then the excess available "type" capacity was assigned to the families proportionately so that each family had the same expected runout time.

Finally, if the sum of the preliminary run quantities of all families in a type exceeded that type's available capacity, the family scheduling subsystem reduced the production quantity of each scheduled family in a type so that equations (2.15) and (2.16) held.

$$RQ_j^* = r_i\left(R_i + O_i\right)RQ_j / \sum_j RQ_j \tag{2.15}$$

$$\sum_j RQ_j^* = r_i\left(R_i + O_i\right)\sum_j RQ_j / \sum_j RQ_j = r_i\left(R_i + O_i\right) \tag{2.16}$$

Again these expressions reflect an approach designed to proportionately scale down individual family production quantities.

The Item Scheduling Subsystem

The item scheduling subsystem used the product family schedules created by the family scheduling subsystem as inputs, and then created the final item production quantity schedule such that the sum of all item quantities in a family equaled the previously assigned family quantity. Similar to the family scheduling subsystem, the item subsystem focused on developing a production schedule for the first month in the planning horizon and it also relied on one or two basic logical formulas rather than a mathematical programming model to produce its schedules.

The item scheduling system employed an approach designed to equalize the expected runout times of all items in a family, subject to satisfying the overstock limits on each item and the overall capacity allocation of a family. The purpose of this approach was to maximize the time until a family had to be rescheduled once a production run of a family took place (*i.e.*, it minimized family changeover costs). Hax and Meal instituted this scheduling approach for the tire manufacturer by using equation (2.17) to develop item schedules.

$$RQ_{kj}{}^{*} = \left\{ F_{kj} \left[RQ_{j}^{*} + \sum_{k} \left(A_{kj} - S_{kj} \right) \right] / \sum_{k} F_{kj} \right\} + S_{kj} - A_{kj} \tag{2.17}$$

where:

$RQ_{kj}{}^{*}$ = desired run quantity for k^{th} Item belonging to Family j

F_{kj} = forecast demand rate for k^{th} Item

$RQ_{j}{}^{*} = \sum_{k} RQ_{kj}^{*}$ = Family run quantity, from the Family Scheduling System

A_{kj} = available inventory of the kth Item

S_{kj} =desired safety stock of the kth Item at the end of the schedule interval

Equation (2.17) created new runout time (2.18) for each item k, and Hax and Meal noted that by (2.17), (2.18) was also equivalent to (2.19) with the runout time for each item k being constant.

$$RT_{k} = \left(RQ_{kj}^{*} + A_{kj} - S_{kj} \right) / F_{kj} \tag{2.18}$$

$$RT_{k} = \left(RQ_{kj}^{*} + \sum_{k} \left(A_{kj} - S_{kj} \right) \right) / \sum_{k} F_{kj} \tag{2.19}$$

Finally, Hax and Meal note that summing over the run quantities calculated by (2.17) for all k items in a family assured that the total production quantity allocated to a family was consumed by its end items as expressed by (2.20)

$$\sum_k RQ^*_{kj} = RQ^*_j \tag{2.20}$$

There were several final tests (for negative item values and item overstock limit violations) and subsequent minor adjustments to item run quantities that occurred to complete the process. However, this summarizes the core activities of the item scheduling system, the final subsystem in the Hax - Meal HPP.

2.2.5 Summary

The Hax-Meal system implemented for a US tire manufacturer remains a landmark case in the field of hierarchical production planning even twenty-five years after its original implementation. This system facilitated integrated decision-making across the corporate and plant levels with higher level, more strategic and tactical decisions generally occurring at the corporate level and more operational, as well as tactical, decisions taking place at the plant level. The design of the system brought this structure to the decision-making process.

The system developed plans which evaluated production and distribution plans network-wide from both a longer term planning perspective and a short run scheduling perspective. Each major subsystem in the HPP provided constraints and guidance to the planning and scheduling activities at the next level down in the planning hierarchy. Finally, the Hax-Meal HPP provided an integrated approach around which the firm could plan and execute with a level of coordination and system-wide perspective that had never previously existed.

2.3 Case Study: An HPP Implemented By A Chemical Fertilizer Manufacturer[3]

In the mid-1970's, a chemical manufacturer and distributor, the Agrico Chemical company, implemented an integrated production, distribution and inventory modeling system to provide decision support for both longer run strategic and tactical issues, and for shorter term operational issues. A multi-discipline group consisting of company representatives and several management scientists from the academic community including Fred Glover and Darwin Klingman led this effort. Although the system's originators did not formally describe their system as an HPP implementation, we will observe that in fact Agrico's system was a network-wide planning system that addressed a hierarchy of planning problems.

At the time of this implementation, Agrico produced eight primary chemical products in four manufacturing plants. Although it manufactured these products throughout the year, the company had highly seasonal demand with up to one-third of annual demand occurring within a four week timespan. The firm operated 78 distribution centers which shipped products to approximately 2000

[3] This section draws heavily upon the reports of this case study found in Glover et. al. (1979). The description of the firm's background and implementation approach is based upon this article.

customers across the US. Transportation between plants and distribution centers took the form of pipeline, barge and rail moves, while trucks were used to make deliveries to customers. As the scope and complexity of the firm's operations continued to expand, the system's developers describe that it had become increasingly difficult for management to address a multitude of production and distribution decisions. Glover et. al. cite the following questions as illustrative of the key issues for which Agrico's management sought improved decision support.

- Where should products be shipped?
- Which demand should be met?
- Which company-owned distribution centers should be used?
- Where should new distribution centers be leased and what should be their size?
- Where and how much product exchange should be made?
- How many railroad cars should be leased and purchased?
- What level of customer service provides the most cost-effective results?
- Which plants or distribution centers should service which clients?

A quick review of these questions reveals a range of operational, tactical and strategic issues. We will now examine the system developed to address these questions in further detail.

2.3.1 Case Problem, Objective And Additional Background

The highly seasonal nature of Agrico's business had continued to place substantial interrelated pressures on the manufacturing and distribution operations of Agrico. These pressures began to manifest themselves in several ways including service fill rate problems and rapidly escalating distribution costs. These pressures prompted Agrico's management to decide that by developing what it termed an integrated PDI ("production, distribution and inventory") planning system, it could better coordinate the activities and operations of the three major decision-making units of its business. Therefore, the company commissioned the development of such a system with the objective of better integrating the efforts of the following three Agrico business segments:

(1) the Supply Segment, responsible for production, purchasing, and product exchange activities with other chemical corporations (coproducers);

(2) the Storage and Customer Distribution Segment, responsible for sizing and locating bulk distribution centers; and

(3) the Demand Segment, responsible for customer demand throughout the eastern two-thirds of the United States, and locations where the product had to be supplied to coproducers due to product exchange agreements.

2.3.2 Overview Of Hierarchical Planning System

Agrico's approach to creating its PDI system centered on developing one major mathematical optimization model which became the centerpiece, and "linking" element of its entire planning system. This network algorithm based optimization model performed multiple functions for both long run and short run planning purposes. Planners who used this model would activate different components, variables and constraints of the model depending upon their particular planning application.

The PDI system developers (*i.e.*, Glover et. al.) linked multiple other existing Agrico decision support systems (e.g., a forecasting system) with the new optimization model to create an overall system. Additionally, Agrico over time also created several additional new system components that it integrated into the PDI system. This approach used by Agrico in developing an integrated HPP system (*i.e.*, build a major optimization model as a centerpiece around which to link existing and other new system components) represents another fairly common strategy used in building HPP systems. A large scale, integrated network-wide production and distribution planning optimization model such as Agrico's touches so many points in a firm's supply chain that it can propel the development of a full scale HPP system. Figure 2.2 illustrates the key components, inputs and outputs of the PDI system. The LCD model, the least cost distribution model depicted in the figure, is the optimization model developed as the core component of the PDI system.

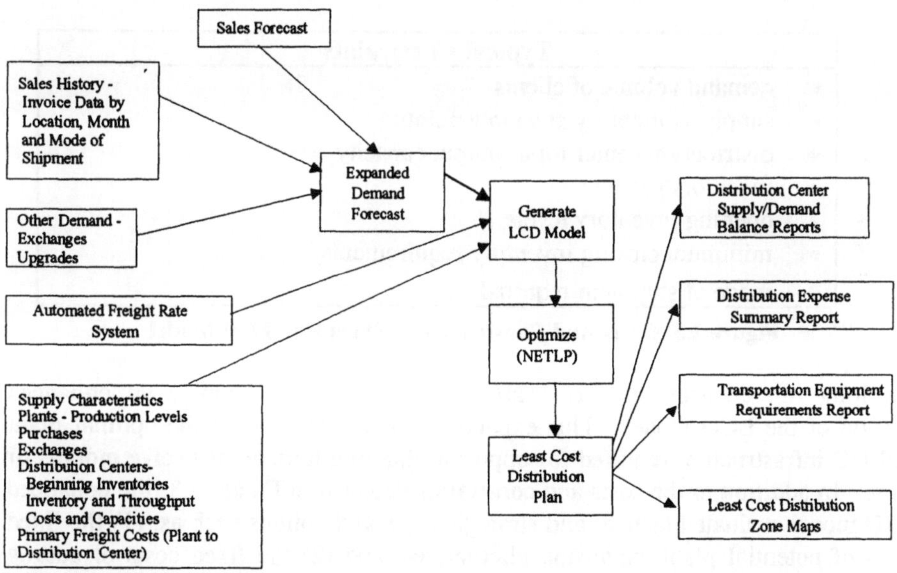

- Reproduced from Glover et. al., 1979.

Figure 2.2 The LCD Model And Other Key Components Of The PDI System

The LCD (Least Cost Distribution) Model

As noted, the LCD model was a network based optimization algorithm which provided very powerful solution capabilities in terms of problem size by existing standards at the time of its implementation.[4] Glover et. al. designed the initial LCD model as a multi-period model with a 12 month rolling horizon and monthly time buckets. Therefore, the model could address both near term planning issues (e.g., for the next quarter) as well as tactical issues (e.g., annual planning problems).

For short run purposes, planners would assume that Agrico's infrastructure, namely its plant and DC network, was fixed open. Then they would employ the LCD model to develop an integrated short run supply/production and distribution plan using this existing infrastructure. Figure 2.3 depicts the typical costs and constraints included in short run planning sessions using the LCD model.

Typical Costs
• variable production costs
• primary costs to move product to distribution center storage
• secondary costs to move product to the client
• variable costs associated with holding inventory
• variable distribution center throughput costs

Typical Constraints
• demand volume of clients
• supply availability at supply points
• distribution center input-output capacity
• inventory capacity
• opening inventory levels
• minimum closing inventory requirements
• mode of shipment required

Figure 2.3 Costs And Constraints In Short Run LCD Model

For long run decision support, Agrico's planners would employ an enhanced version of the LCD model. This expanded version determined the optimal plant and DC infrastructure required to support a planning horizon of twelve months or more. In addition to the costs and constraints depicted in Figure 2.3, the enhanced LCD model evaluated tactical and strategic costs and options such as: (1) the fixed costs of potential plant expansion alternatives, and (2) the fixed costs of adding additional DC's to the existing network.

How The LCD Model Fit Into The Overall Planning System

The overview of the PDI system shown in Figure 2.2 depicts how Agrico integrated the LCD model with its existing and new supply chain planning components. Existing historical sales databases with detailed sales data were used in conjunction

---

[4] See Glover et. al. (1979) for a detailed description of the LCD model algorithm.

with a higher level forecasting system to generate demand forecasts to input to the LCD model. Eventually the need for more accurate freight rate data spawned the development of an automated freight rating system. This system was then integrated to feed the LCD model. Over time, Agrico also expanded its decision support system providing supply data about plants and DCs' costs, capacities, inventory levels and so on. This supply information also fed the LCD modeling system.

On the output side, the LCD model generated a series of solution reports addressing a hierarchy of planning decisions. These solution reports ranged from the LCD zone map reports which Agrico used daily to control the distribution of the placement of orders (*i.e.*, order fulfillment), to other solution reports not shown in Figure 2.2 which directed long term plant and DC capacity expansion decisions.

Figure 2.4 illustrates Agrico's initial project design for its PDI system. The design highlights the flow of information and communications throughout the PDI system and through the Agrico organization. In particular, note the feedback loop (entitled "revisions") shown in Figure 2.4. This represents a critical component of a truly integrated system. The PDI developers conceived and then implemented a system where business managers, production planners, transportation planners, the customer service group and others would both: (1) regularly use the LCD model for planning purposes, and even more importantly, (2) regularly revise planning inputs as appropriate based upon real world operating conditions. As a rule, the successful implementation of a large scale planning and scheduling system depends upon these two attributes: (1) planners must use the system as a standard business practice, and (2) planners must have the flexibility to provide feedback and revisions to the planning system's methodology and outputs.

2.3.3 Summary

The implementation of the PDI system produced dramatic benefits for Agrico Chemical Company. The system's designers reported decreases in annual operating costs of between 10% and 15% in the first four years after the implementation of the system. Although not an HPP implementation in the classical sense (*i.e.*, the Hax-Meal model), the PDI system represented a hierarchical planning system in its own right. The PDI addressed planning needs ranging from the day-to-day operational through annual and even strategic decision support. As such, a wide variety of Agrico's functional areas and levels of management either directly or indirectly used the system. Further, the system's developers reported that their key implementation strategy consisted of placing heavy emphasis on involving numerous representatives from all three business segments in developing the PDI system. Thus, once they implemented the PDI system, a substantial "vested" constituency already existed within Agrico who supported the use of this system for planning purposes.

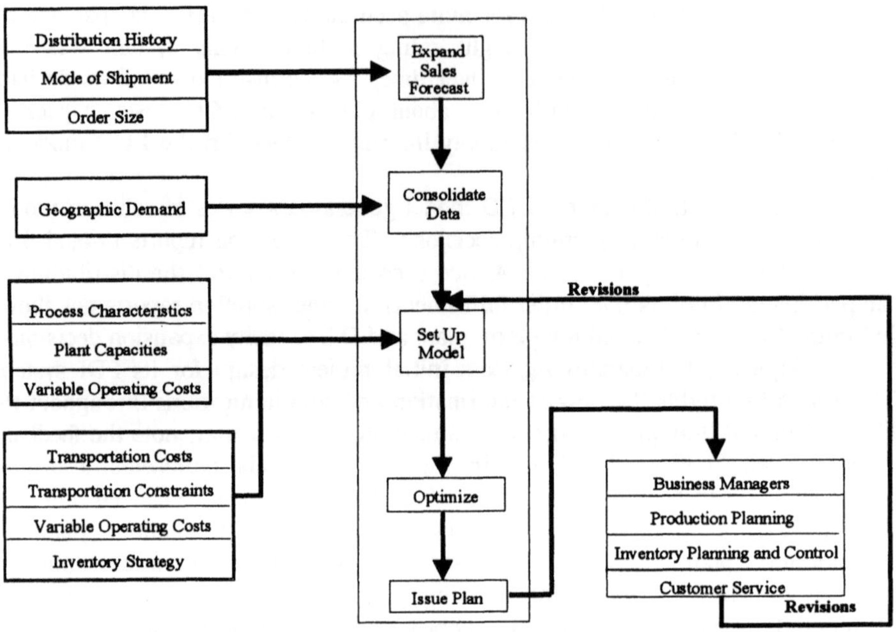

- Reproduced from Glover et. al., 1979

Figure 2.4 Initial Project Design For Agrico's PDI System

Finally, the PDI system produced several other benefits consistently found in successful HPP implementations. Glover et. al. describe that overall information flow and communications significantly improved within Agrico with the onset of this system. If one considers the broad set of inputs to, outputs from, and users of the LCD model, intuitively it makes sense that communications would improve as described. A "centerpiece" model or system that supports many user groups typically produces this effect. Additionally, Glover et. al. reported that the initial development of the PDI system eventually stimulated the creation and/or enhancement of other system components. This resulted because as managers began to utilize the PDI system, the need for both better data and missing decision support data became clear. This again represents a typical evolution of a HPP planning system. Finally, Glover et. al. also cite that once they successfully implemented the initial version of the LCD model, a process of continuous enhancements to this modeling system evolved as the system's designers attempted to accommodate the ever increasing informational demands of management. These enhancements included such features as providing : (1) the ability to model additional cost factors, (2) the flexibility to evaluate production and purchase "ranges" rather than fixed levels, and (3) more sophisticated capabilities to evaluate seasonal demand patterns, production capacities, inventory investment costs, transportation equipment capacities, and so on. Again this represents a typical characteristic of a large scale HPP implementation.

2.4 Case Study: An HPP Implemented By A Ceramic Tile Manufacturer [5]

In the late 1980's, American Olean Tile Company (AO), a ceramic tile manufacturer, implemented an integrated hierarchical production and distribution planning system. The firm used this system to provide decision support and direction in its strategic, tactical and operational planning activities. Similar to the HPP implementations of the tire and chemical fertilizer manufacturers previously reviewed, American Olean's system evaluated its entire manufacturing and distribution network. Thus, AO's HPP produced planning solutions which directed manufacturing and distribution operations in an integrated fashion.

2.4.1 Case Problem, Objective And Additional Background

The American Olean Tile Company manufactured a wide variety of ceramic tile products for walls, floors and elaborate mural designs. At the time it implemented its HPP system, AO operated eight factories located across the US from New York to California that supplied approximately 120 sales distribution points (SDP's) which consisted of a combination of sales territories and company-owned warehouses. These factories utilized several different production processes, all of which began with a crushing and milling procedure, and which eventually lead to the firing of tile in large kilns. AO produced three basic lines of tile products: (1) glazed tile, (2) ceramic mosaics, and (3) quarry tile. The quarry division operated four factories at three locations (Pennsylvania, Kentucky and California), while the glazed and ceramic mosaic divisions each had two manufacturing sites, respectively.

During the early and mid-1980's, American Olean's distribution network had expanded quite rapidly and the firm's management expected this trend to continue into the 1990's. As the firm's network grew, the lack of coordination which existed between manufacturing planning and distribution planning had become an increasingly significant problem for the firm. For example, AO frequently found itself unnecessarily shipping finished goods inventory back and forth between distribution centers because manufacturing and distribution planned their operations independently. This situation eventually prompted AO's management to sponsor a modeling program to address this planning problem. Initially the program was chartered simply to integrate the annual production and distribution planning processes. However, shortly after the project began, it became clear that AO would not reap the full benefits from integrated annual plant, product and distribution assignments if it did not also assure that alignment existed between short-term scheduling and inventory control decisions, and annual plans. Thus, AO's management commissioned the development of a full-scale hierarchical production and distribution planning system. The objective of implementing this system was to improve the integration of annual production and distribution planning, short-term scheduling and inventory control.

---

[5] Some of the material in this section is adapted and expanded from Liberatore and Miller (1985).

The development of AO's HPP system began with its quarry division, which had the largest manufacturing and distribution network of its three divisions. Eventually, the system was implemented across all three divisions. Importantly, each of the divisions' plants could manufacture most of the major products of their respective divisions. Further, the cost structures and the manufacturing technology varied quite significantly from plant to plant within each respective division. Therefore, significant manufacturing and distribution cost trade-offs existed within each division, and AO needed a system that would simultaneously consider these trade-offs in determining how to most efficiently utilize the firm's infrastructure.

2.4.2 American Olean's Hierarchical Planning System

The designers of American Olean's HPP system implemented it with the intent that their system would support the firm's network-wide strategic, tactical and operational planning activities. The approach used by AO closely followed the generic HPP framework outlined in Chapter 1, and Figure 2.5 illustrates this framework. As reviewed earlier, decisions made at each planning level in the hierarchy provide direction and place constraints upon the level(s) below it and/or provide feedback to the level(s) above it. The focus of AO's management rested primarily at the tactical and operational levels (and our discussion will now also focus on these levels). Nevertheless, from the outset AO managers envisioned using their HPP for strategic planning also. This created the need to construct a system with the full-scale hierarchy shown in Figure 2.5.

The Pre-HPP Production And Distribution Planning System At AO

The benefits of AO's HPP system can perhaps best be understood by first briefly reviewing the firm's annual planning and short run production and distribution scheduling system prior to the HPP implementation. Figure 2.6 displays AO's pre-HPP annual (tactical) and operational planning and scheduling decision support system. The firm's annual planning basically consisted of three independent activities: (1) the development of high level divisional product line forecasts by the marketing department, (2) the development of a plant level production plan for each plant by the manufacturing department, and (3) the development of a distribution sourcing plan by the distribution department. These three activities were not integrated (e.g., the manufacturing plan was not driven by the annual forecast) and were primarily manual exercises. Any "cost-based" decisions in the manufacturing or distribution planning activities considered only the costs of the individual respective functional area.

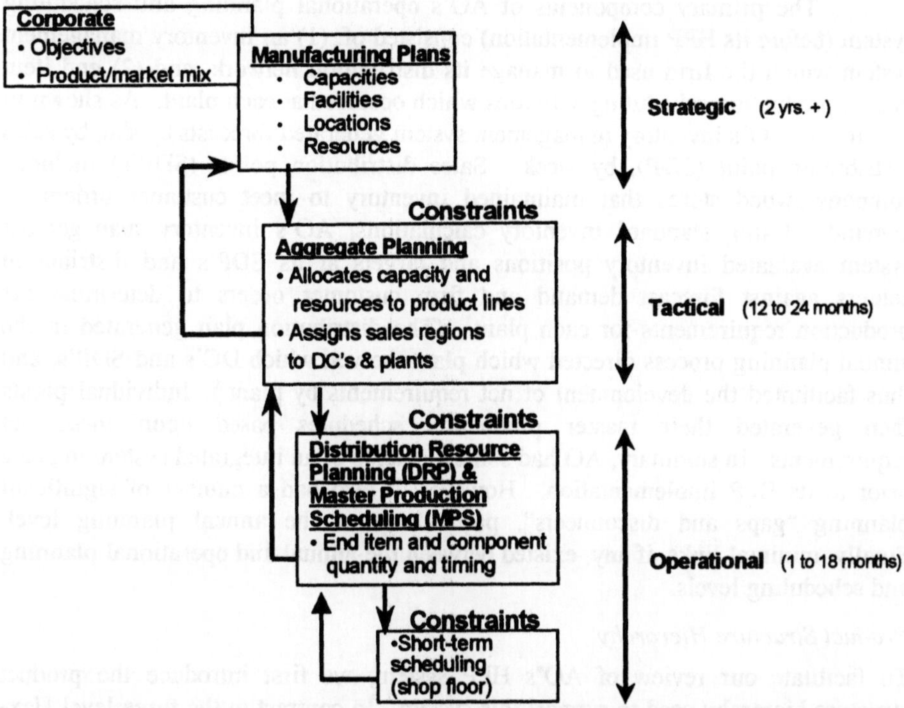

Figure 2.5 American Olean's Hierarchical Production Planning Framework

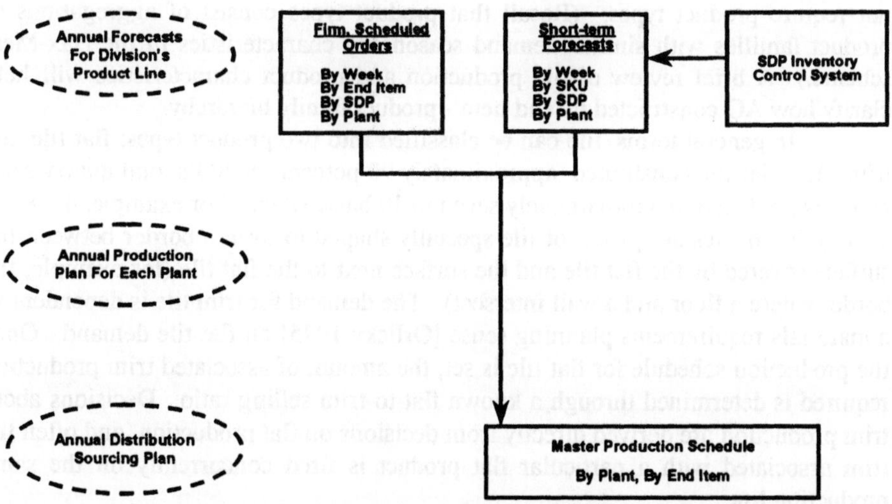

Figure 2.6 Production And Distribution Planning System-Pre-HPP Implementation

The primary components of AO's operational planning and scheduling system (before its HPP implementation) consisted of: (1) an inventory management system which the firm used to manage its distribution network, and (2) end item master production scheduling activities which occurred at each plant. As shown in Figure 2.6, AO's inventory management system generated forecasts by sku, by sales distribution point (SDP), by week. Sales distribution points (SDP's) included company-owned stores that maintained inventory to meet customer orders on demand. Using standard inventory calculations, AO's inventory management system evaluated inventory positions and targets at its SDP's and distribution centers against forecast demand and firm customer orders to determine net production requirements for each plant. (The distribution plan generated in the annual planning process directed which plants served which DC's and SDP's, and thus facilitated the development of net requirements by plant.) Individual plants then generated their master production schedules based upon these net requirements. In summary, AO had some elements of an integrated system in place prior to its HPP implementation. However, it also had a number of significant planning "gaps and disconnects", particularly at the annual planning level. Finally, minimal links, if any, existed between the annual and operational planning and scheduling levels.

Product Structure Hierarchy

To facilitate our review of AO's HPP system, we first introduce the product structure hierarchy used to support this system. In contrast to the three level Hax-Meal product hierarchy used by the tire manufacturer, AO employed a simpler two level product hierarchy consisting of end items and product families.[6] AO chose to explicitly forecast demand seasonality at the product family level, and therefore, did not require product types. (Recall that product types consist of aggregations of product families with similar demand seasonality characteristics in the Hax-Meal scheme.) A brief review of tile production and product characteristics will help clarify how AO constructed its end item - product family hierarchy.

In general terms, tile can be classified into two product types: flat tile and trim tile. Flat tile constituted approximately 90 percent of AO's total quarry sales and was produced in approximately seven to 10 basic shapes (for example, 4"x8" or 6"x6"). Trim tiles are pieces of tile specially shaped to form a border between the surface covered by the flat tile and the surface next to the flat tile (for example, the border where a floor and a wall intersect). The demand for trim tile is dependent in a materials requirements planning sense [Orlicky 1975] on flat tile demand. Once the production schedule for flat tile is set, the amount of associated trim production required is determined through a known flat-to-trim selling ratio. Decisions about trim production are derived directly from decisions on flat production, and often the trim associated with a particular flat product is fired concurrently on the same production line.

---

[6] In reality, both systems' product hierarchies consisted of more than just three and two levels respectively. For example, in both systems, product type (Hax-Meal) and product family (AO) forecasts aggregated up to overall division and corporate forecasts.

AO's flat tile production process itself made it logical to further condense end items into major product families. In addition to a basic shape and color, flat tile is made in several different surfaces (for example, regular and abrasive), and in several variations of the basic color (for example, Grey and Grey Flash). However, tiles made from one basic flat shape in one color all require very similar raw materials and have virtually identical manufacturing costs and capacity constraints. Therefore, several to many major flat tile end items could be aggregated into one major product family with no impact on planning accuracy. In the case of AO's quarry division, the aggregation process resulted in the formation of 10 major product families encompassing over 98 percent of total quarry sales. A similar aggregation process for the glazed and ceramic tile divisions resulted in the creation of approximately 35 and 12 product families, respectively.

The HPP Production And Distribution Planning System At AO

AO fully implemented its hierarchical production and distribution planning system in about two years in total. The designers of AO's HPP adopted an approach that contained some elements of both the Hax-Meal and the Agrico Chemical methodologies. Similar to the Hax-Meal tire manufacturer implementation, AO did construct a formal HPP that linked numerous decision support components providing both tactical and operational planning and scheduling support. Similar to the Agrico implementation, AO's designers employed one major new component, an annual production and distribution planning optimization model, as the core module around which to develop and enhance the entire HPP system. And, similar to both the tire manufacturer's and chemical fertilizer's HPP's, AO chose to build a system that could provide integrated manufacturing and distribution planning support for its entire network. In this section, we will review the entire HPP at a summary level, while two subsequent sections will provide detailed descriptions of the HPP's two optimization models (*i.e.*, the "core" tactical/annual production and distribution model and the operational plant production scheduling model).

Figure 2.7 summarizes the framework used for the hierarchical integration of the production and distribution planning and scheduling activities of each AO tile division. The planning process began with an annual (or rolling twelve to eighteen month) sales forecast for each division's sales expressed in square feet of tile. The director of market planning, in conjunction with other senior managers, generated this sales projection based upon a combination of economic trends and specific division related market developments. AO planners would then allocate this division level forecast to each product family, and then apportion each family forecast to the SDP's based upon the ratio of the SDP's annual sales of the product family to the total sales of the product family during the previous twelve months. At the time of the system's implementation, AO conducted a study which revealed that its sales distribution mix remained essentially constant in the short run (12-18 months). Therefore, this ratio based forecast dissaggregation method worked quite well for AO.

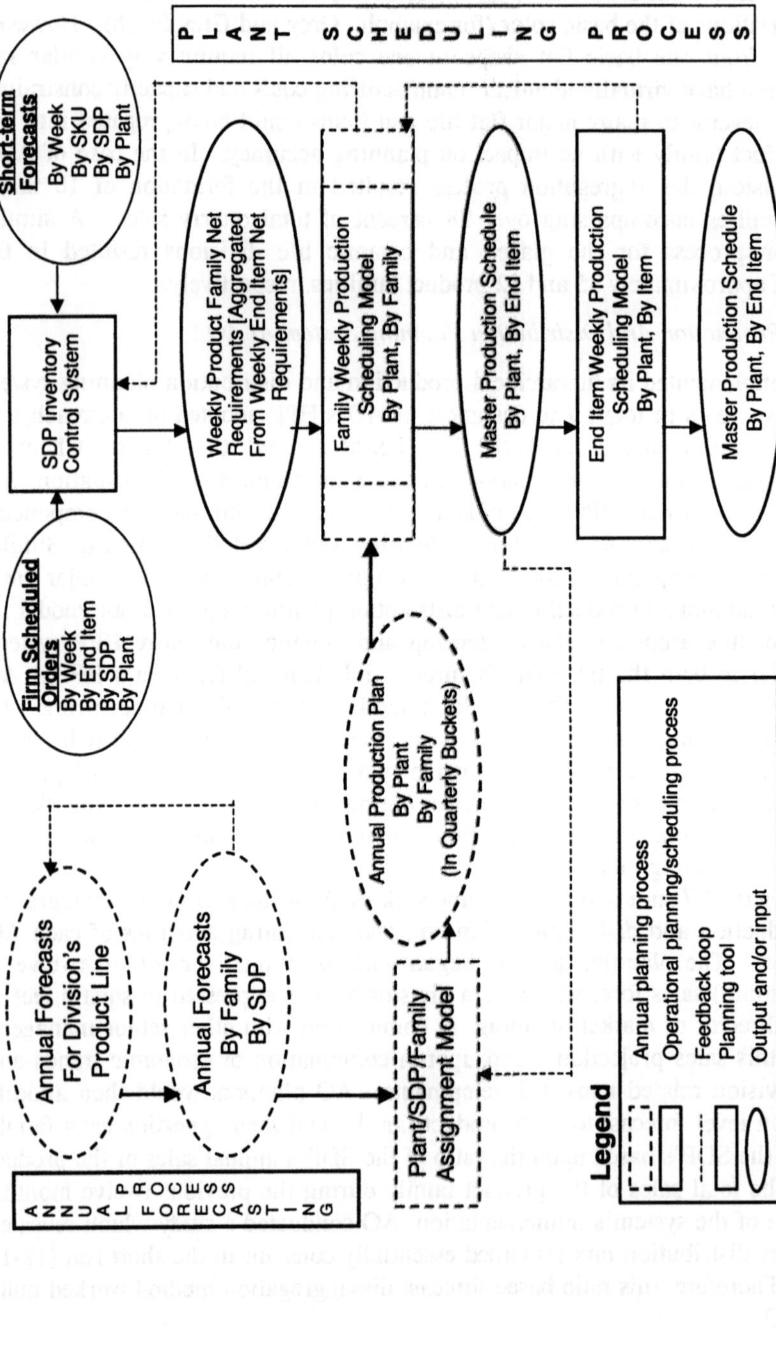

Figure 2.7 Production And Distribution Planning System-Post-HPP Implementation

It is important to note that AO over time enhanced its forecasting process. Specifically, it began to predict annual sales of each product family using an exponential smoothing based forecasting and tracking system. As depicted in Figure 2.7, this enabled AO to institute a hierarchical forecasting feedback system that compared total division and product family forecasts generated by: (1) a top down - division to product family approach, and (2) a bottom up - product family to division approach. As part of this system, AO established criteria to identify and reconcile significant differences between top down and bottom up generated forecasts. This represents a good example of the type of evolution that typically occurs during and after HPP implementations.

The next component in AO's HPP system consisted of a tactical (annual) mathematical optimization model denoted the plant/SDP/family assignment model (hereafter called the assignment model). The twelve to eighteen month forecasts generated by the annual product family forecasting system provided the demand constraints that drove the assignment model. Briefly, the model minimized total variable production, freight and inventory carrying costs while meeting SDP forecast demands and minimum SDP and DC inventory targets within available plant capacities. The principal output of this model was an integrated production and distribution plan for AO's plants, DC's and SDP's. Specifically, the model would create an annual distribution plan that assigned the SDP's to DC's and plants (*i.e.*, a sourcing plan) and established planned interplant and inter-DC shipping plans. Additionally, the assignment model generated an annual production plan for each plant, by quarter, with planned operating rates (*i.e.*, capacity utilization rates) and planned production mixes by product family. Finally, the assignment model provided an inventory plan over the planning horizon with projected inventory levels by product family, by quarter.

As shown in Figure 2.7, the product family weekly scheduling model represented the next component of the annual/tactical planning system. In fact, this model provided decision support for both the annual planning process as well as for the short run scheduling of operations. Briefly, the family scheduling model minimized the sum of variable production costs, changeover costs and inventory carrying costs and it was really a short run, operational plant scheduling model by design. The model could evaluate a flexible planning horizon (anywhere from one quarter to one year) in weekly time buckets (e.g., 52 for an annual planning horizon). The family scheduling model assisted the annual planning process by evaluating the feasibility of the annual production plan generated by the assignment model for each plant. Specifically, AO planners could input the quarterly product family production plans generated for a plant by the assignment model, and evaluate how the family scheduling model would actually schedule these production "forecasts" by week. Because the family scheduling model included changeovers and because it evaluated product family production plans in weekly time buckets (which is how the AO plants scheduled product families), the model could determine whether an individual plant could really schedule and produce the production plan suggested by the annual assignment model. Imbedding this linkage and feedback between the tactical and operational planning systems as part of the "annual (tactical)" planning process represented a critical element in AO's HPP implementation. Finally, we should note that the actual process of creating weekly, product family production forecasts from the annual assignment model's four quarterly family production plans involved planners using weekly seasonal product family indices. Essentially, planners would disaggregate quarterly plans into weekly buckets based on these indices and make any additional adjustments as needed. This completes the overview of the tactical or annual components of AO's HPP system. We now turn to the operational planning and scheduling part of this system.

As previously noted, AO had in place an inventory management and forecasting system which directed distribution network operations (e.g., inventory replenishment and transport delivery activities) prior to the implementation of its HPP system. Thus, AO simply incorporated its inventory management system into the overall HPP scheme. As previously described, the inventory management system generated forecasts by sku, by sales distribution point, by week, and employing standard inventory methodologies, it evaluated inventory positions and targets at AO's SDP's and DC's against forecast demand and firm orders to determine the net production requirements for each plant. The sourcing assignments (*i.e.*, which plants supplied which DC's and SDP's) did change as a result of the implementation of AO's assignment model. However, other than implementing these sourcing assignment changes, AO did not modify how it used its inventory management system.

As noted, the inventory management system generated production net requirements by week, at the end item level for each plant. In order to schedule these production requirements at the product family level using the family weekly scheduling model, AO planners aggregated the end item net requirements into product family net requirements, as Figure 2.7 illustrates. Then the plant family weekly production scheduling model would use these family net requirements as the demand constraints to drive a model solution (*i.e.*, a product family weekly master production schedule). Note in Figure 2.7 that a feedback loop flows from the family scheduling model back to the short term end item forecast (and net requirements) generated by the inventory management system. In the event that the family weekly plant scheduling model could not produce a solution that met all of the net requirements demand, this infeasibility would be communicated back. At this point, planners would have to adjust the forecast or requested net family production requirements in order to facilitate a feasible family master production schedule at a plant. [7]

The final major step in the operational part of AO's HPP planning and scheduling system consisted of translating the product family weekly master production schedule into an end item weekly master production schedule. Changeovers on a production line from one item to another within a product family required only a fifteen minute washup at a break, and did not result in any lost

---

[7] If customer or SDP demand requests had to be backordered as a result of this process, this "feedback" information also would be communicated.

production time or capacity. Thus, AO schedulers could develop the end item production schedules quite effectively using simple spreadsheet model based calculations. It is important to note that a feasible product family master production schedule can in fact contain infeasibilities which only become visible when evaluated at the end item level. For this reason, Figure 2.7 shows a feedback loop flowing from the end item master production schedule back to the weekly product family scheduling model. In practice, this occurred quite infrequently at AO. This completes our review of the network-wide HPP system implemented by American Olean Tile Company. We now will explore the two optimization components of this system in further detail.

The Annual Plant/SDP/Family Assignment Model [8]

AO's original annual model consisted of a single-period transportation-type linear-programming (LP) formulation. As noted, the principal output of the model was an integrated production and distribution plan that minimized total variable production and freight costs while meeting SDP forecasted product demands within available plant capacities. Once AO's manufacturing operations began to use these models as primary input to the annual and tactical capacity planning process, planners quickly recognized the need to create multi-period models with inventory flows and planning targets. The multi-period models improved AO's ability to recognize and plan for overcapacitated short-run periods (within a year) because they could capture by quarter such factors as demand seasonality and changes in production line capacities. Thus, AO revised the annual model into a dynamic LP model that included more parameters, such as inventory carrying costs per period and inventory targets by product, by period for all intermediate and final stocking locations. Table 2.5 displays the notation for the dynamic plant/SDP/family assignment model, and equations (2.21) – (2.28) illustrate the model.

Table 2.5 Notation For The Plant/SDP/Family Assignment Model

$p =$	index on the plant
$i =$	index on the product family
$j =$	index on the production line
$k =$	index on the sales distribution point
$t =$	index on the production period (in units of 1 quarter)
$c_{pijt} =$	the variable unit cost per square foot of producing family i at plant p on line j during period t
$v_{pikt} =$	the unit cost per square foot to ship family i from plant p to sales distribution point k during period t
$w_{pip't} =$	the unit cost per square foot to ship to family i from plant p to plant p' (where $p \neq p'$) during period t
$h_{pit} =$	the unit cost per square foot of carrying family i in inventory at plant p from period t to period t+1
$d_{ikt} =$	square feet of family i demanded at sales distribution point k during period t

[8] The material in this section is adapted and expanded from De Matta and Miller (1993).

Table 2.5 Notation For The Plant/SDP/Family Assignment Model (continued)

o_{pit} =	**the end of period minimum inventory objective or target (in square feet) for family *i* at plant *p* during period *t***
z_{pijt} =	**total potential production capacity (in square feet) to produce family *i* at plant *p* on production line *j* during period *t***
Define the decision variables	
Y_{pijt} =	**square feet of product family *i* produced at plant *p* on production line *j* during period *t***
S_{pikt} =	**square feet of family *i* shipped from plant *p* to sales distribution point *k* during period *t***
$T_{pip't}$ =	**square feet of family *i* shipped from plant *p* to plant *p'* during period *t* (where $p \neq p'$)**
I_{pit} =	**square feet of family *i* in inventory at plant *p* at the end of period *t***

The Plant/SDP/Family Assignment Model

Minimize:
$$\sum_p \sum_i \sum_j \sum_t Y_{pijt} c_{pijt} + \sum_p \sum_i \sum_k \sum_t S_{pikt} v_{pikt}$$
$$+ \sum_p \sum_i \sum_{\substack{p' \\ p' \neq p}} \sum_t T_{pip't} w_{pip't}$$
$$+ \sum_p \sum_i \sum_t I_{pit} h_{pit} \qquad (2.21)$$

Subject to:
$$\sum_p S_{pikt} = d_{ikt} \quad \text{for all } i, k, t \qquad (2.22)$$

$$I_{pi(t-1)} + \sum_j Y_{pijt} + \sum_{\substack{p' \\ p' \neq p}} T_{p'ipt} - \sum_{\substack{p' \\ p' \neq p}} T_{pip't}$$
$$- \sum_k S_{pikt} - I_{pit} \geq o_{pit}, \quad \text{for all } p, i, t \qquad (2.23)$$

$$\sum_i (1/z_{pijt}) Y_{pijt} \leq 1, \quad \text{for all } p, j, t \qquad (2.24)$$

$$Y_{pijt} \geq 0, \quad \text{for all } p, i, j, t \qquad (2.25)$$

$$S_{pikt} \geq 0, \quad \text{for all } p, i, k, t \qquad (2.26)$$

$$T_{pip't} \geq 0, \text{ for all } p, i, p', t \tag{2.27}$$

$$I_{pit} \geq 0, \text{ for all } p, i, t \tag{2.28}$$

Equations (2.22), (2.23) and (2.24) represent respectively the demand, the inventory target (*i.e.*, the objective) and the capacity constraints. The capacity constraints (2.24) include a multiplier ($1/z_{ijpt}$). This serves to adjust for the fact that each of the product families that can be produced on a given production line may have unique capacity levels (or equivalently, production rates). Note that this model excluded fixed manufacturing costs because AO's standard planning approach generally assumed that the firm would operate all existing plants over the planning horizon.

AO planners developed several versions of the annual model, all of which evaluated the same major costs, but with minor variations. The variations ranged from the LP formulation (2.21) - (2.28), to several modified formulations that included fictitious production capacities with associated pricing schemes on the costs of utilizing them. The pricing (costing) schemes in these models ensured that, in periods where forecasted demand exceeded projected available capacity and inventory, capacity was allotted to product families in accordance with the priorities set by sales and marketing management. This represented an attempt to minimize the cost of potential lost sales. The modified LPs helped AO planners to obtain feasible solutions when corporate forecasts exceeded actual plant capacities, and more importantly, to identify effective capacity allocation plans for projected over-capacitated periods.

Briefly, AO constructed models with dummy production lines whose capacities exceeded corporate forecasts by factors of 10 times or more (that is, essentially infinite capacity). Unit costs to produce product families on these lines were set much higher than the unit costs on real production lines. Thus, when cost was minimized, dummy lines were utilized as a last resort. However, on these lines, planners constructed and manipulated the unit production costs assigned to particular product families so that they could control the order by which product families were assigned to real and dummy production lines. Product families judged of lower priority by sales and marketing management were given relatively lower unit costs on the dummy lines than product families judged of higher priority. Therefore, if some production had to be assigned to dummy production lines in order to generate a feasible model solution, those product families of lower priority would be assigned first to the dummy production lines. This modeling approach proved invaluable for the production planners who had limited time to develop corporate production plans.

AO also developed and maintained several mixed-integer models that addressed such factors as minimum run lengths on production lines and distribution service requirements, such as specific sales regions that required single plant sourcing. The linear programming formulations of the annual model, however, satisfied the planning requirements in well over 90 percent of the planning cycles. This resulted from several factors. First, as noted, AO generally planned to operate all available facilities over defined planning horizons (for example, 12 to 18 months) and, therefore, typically considered only variable manufacturing or facility costs. Second, continuous formulations generally provided good solutions for allocating capacity and product mixes to plants and for establishing unique plant/product family/SDP sourcing assignments.

The Family Weekly Plant Production Scheduling Model

The family weekly plant production scheduling optimization model was the last component of the HPP system to be fully implemented. Initially AO's planners developed weekly product family production schedules using simpler spreadsheet based models until the firm developed and then began utilizing optimization models to assist this process. As previously described, the family scheduling model played a dual role in AO's HPP system. Specifically, AO used the model to:

1. assist in evaluating the feasibility of annual production plans generated by the tactical plant/SDP/family assignment model, and
2. provide decision support for the short run operational scheduling of plants.

The model minimized the total of variable production, line changeover and inventory holding costs over the planning horizon while meeting product family demand and inventory requirements. A model solution produced a master production schedule at the family level for a plant. Production rates by family by production line were constant during each scheduling period and backlogging of demand was not permitted. The model itself was a mixed integer linear program, and Table 2.6 displays the notation for the mathematical formulation.

Table 2.6 Notation For Family Weekly Plant Production Scheduling Model

X_{ijt}=	1, if product *i* produced on production line *j* during period *t*; $X_{ijt} = 0$, otherwise
V_{ijt}=	1, if a production changeover to product *i* on line *j* occurs during time *t*; $V_{ijt} = 0$, otherwise
C_{ij} =	the unit cost of producing product *i* on line *j*
P_{ij} =	the production rate per period of product *i* on line *j*
S_{ij} =	the fixed cost to set up product *i* on line *j*
D_{it} =	the demand for product *i* during period *t*
H_i =	the end-of-period holding cost for product *i*
I_{it} =	the inventory of product *i* at the end of period *t*
L_i =	the set of all production lines which can be scheduled to produce product *i*
M_j =	the set of all products which can be scheduled on production line *j*
P =	the number of products to be scheduled
L =	the number of production lines available
N =	the number of periods in the planning horizon
t =	the time period, expressed in units of one week

The family scheduling model was formulated as follows:

Minimize:
$$\sum_{t=1}^{n}\sum_{j\in L_i}\sum_{i=1}^{P} C_{ij}P_{ij}X_{ijt} + S_{ij}V_{ijt} + \sum_{t=1}^{N}\sum_{i=1}^{P} H_i I_{it} \tag{2.29}$$

Subject to:
$$I_{it} \geq 0, \qquad i=1, 2 \ldots, P,\ t=1, 2 \ldots, N \tag{2.30}$$

$$\sum_{i\in M_j} X_{ijt} \leq 1, \ j=1, 2 \ldots, L,\ t=1, 2 \ldots, N \tag{2.31}$$

$$V_{ijt} \geq X_{ijt} - X_{ij,t-1} \qquad i=1, 2 \ldots, P, \quad j \in L_i, \quad t=1, 2 \ldots, N \tag{2.32}$$

$X_{ijt}V_{ijt}$ binary, where $I_{i0}X_{ij0}$ given as initial conditions

Finally, ending inventory is defined as

$$I_{it} = I_{i0} + \sum_{k=1}^{t}\sum_{j\in L_i} P_{ij}X_{ijk} - \sum_{k=1}^{t} D_{ik} \tag{2.33}$$

$$i=1, 2 \ldots, P,\ t=1, 2 \ldots, N$$

Equation (2.30) prevents backlogging of demand, while equation (2.31) insures that at most one product is scheduled on a given line during each period (an assignment type constraint). Equation (2.32) relates the changeover variable (V_{ijt}) to the production scheduling variables in the current and previous periods (X_{ijt} and $X_{ij,t-1}$ respectively). A changeover to product i on line j occurs during period t if and only if $X_{ijt}=1$ and $X_{ij,t-1}$ is 0. Thus even if the line was idle during period t-1, it is assumed that a changeover occurs to i if it is then scheduled during period t. It is easily shown that equation (2.32) forces V_{ijt} to be 1 only if $X_{ijt}=1$ and $X_{ij,t-1}$ is 0; in all other cases V_{ijt} must be 0. Thus V_{ijt} need not be specified as a binary variable.

2.4.3 Summary

American Olean Tile Company's HPP implementation produced dramatic benefits for the company. The firm reduced annual operating costs by about 10% within a year after implementation. The manufacturing, marketing, sales and distribution groups began to develop integrated tactical and annual plans using the plant/SDP/family assignment model as their unifying decision support tool. The firm experienced improved alignment between annual production plans and short run production schedules. As also observed in the Agrico Chemical case, the implementation of AO's HPP stimulated the development of additional new systems and enhancements to existing systems beyond the original scope of the HPP system design. For example, the need for improved forecasts to drive the annual assignment model eventually led to the development of an enhanced product family and SDP forecasting system.

American Olean also improved its overall planning process in an area that occasionally receives less attention in HPP implementations, namely, strategic planning. In the next several years following the development of its HPP system, AO had to evaluate potential major changes to its network infrastructure (e.g., opening major distribution centers in the western U.S. and major enhancements to its manufacturing capabilities). When these strategic decisions arose, the existence of a well functioning HPP system with a tactical optimization model anchoring its annual planning process facilitated relatively quick and thorough quantitative, as well as qualitative evaluations of these strategic questions.

Finally, AO's implementation of its HPP led to a significant improvement in coummunications across many areas of the firm. As noted, the manufacturing, sales, marketing and distribution groups began to carry out annual/tactical planning in a coordinated approach. A formal process was established whereby these groups jointly updated rolling twelve to eighteen month plans on a quarterly basis. Communications between tactical planning personnel and operating personnel also improved as the common platform of an HPP sparked a more collaborative approach. In particular, the fact that all parties were providing portions of the input to the HPP, and similarly accepting the planning outputs of the various components of the system, contributed to this improvement.

The improvement in communications resulting from major HPP and systems implementations represents a common theme found in the three cases of American Olean Tile, Agrico Chemical and the tire manufacturer studied in this chapter, as well as many other cases. For this reason, we close this chapter with a review of several illustrative techniques used in conjunction with HPP implementations to foster improved communications.

2.5 Communications In A Hierarchical Production Planning Environment [9]

As part of its HPP implementation, American Olean Tile Company also implemented a "communications" program to assure the successful use of its new planning system. The communications program was in large part based on the philosophy that AO had rolling, integrated tactical and operational plans (generated by its HPP) that required constant promotion and visibility throughout its organization. The firm recognized that it needed to employ both formal and informal procedures to update and modify its plans on an ongoing basis. At the informal level, frequent communications took place between key players at the operational and tactical planning levels. These dialogues included both headquarters personnel and field personnel such as production schedulers, plant managers and customer service managers at the plants and DC's, and sales managers and key account representatives in field offices.

[9] Portions of the material in this section are adapted and expanded from Miller and Liberatore (1988).

More formally, AO established a monthly production, sales and inventory review committee to assure a regular communications forum for reviewing tactical and operational plans. (This sort of monthly meeting is probably most commonly known as the Sales And Operations Planning – S&OP – Review Meeting in most companies.)[10] This group had the responsibility of reviewing the latest sales, marketing, production and inventory data to determine whether the current tactical (annual) plan required adjustment. The planning horizons considered by this group began with the short run (four to thirteen weeks) and extended over at least the next twelve to eighteen months. Committee members included vice-presidents and directors from each key function, as well as key day-to-day operations personnel from such areas as production scheduling and customer service.

The formal deliberations of the monthly review committee were facilitated by a monthly update of key components of AO's HPP system. Specifically, the firm would update its rolling tactical/annual forecasts and input these forecasts into the plant/SDP/family assignment model. Results of these modeling exercises would help guide any modifications that the group would make to its rolling tactical business plans.

Disseminating Optimization Model Data Throughout The Organization

In implementing the optimization model components of its HPP, AO's planners were sensitive to the issue that optimization models were generally perceived as "black boxes" by virtually the entire organization. This can lead to considerable difficulties in achieving acceptance of these models by the rest of the organization. Therefore, the planners responsible for the optimization models disseminated as much information as possible throughout the firm concerning exactly what data these models used, and what solution outputs these models produced. Once this information became well known, the firm's acceptance of the optimization modeling based approach increased dramatically. Table 2.7 illustrates the type of internally disseminated, simple production data which AO's planners distributed. This table described the production rates and possibilities of two of AO's plants as modeled by the plant/SDP/family assignment model.

Table 2.7 Production Possibilities Of The Tactical /Annual Assignment Model

		Plant 1		Plant 2		
Product Family	Description	Production Kiln 01 Net Prod Units/wk	Production Kiln 02 Net Prod Units/wk	Production Kiln 03 Net Prod Units/wk	Production Kiln 04 Net Prod Units/wk	Production Kiln 05 Net Prod Units/wk
A	1"x 1"	40,000	30,000	-	25,000	30,000
B	2"x 1"	-	40,000	-	35,000	45,000
C	2"x 2"	60,000	50,000	-	45,000	50,000
D	1" HEX	-	20,000	-	15,000	30,000
E	2" HEX	-	10,000	-	5,000	25,000
F	4"x 4"	70,000	60,000	90,000	55,000	40,000
G	6"x 6"	80,000	-	100,000	-	45,000
H	8"x 8"	70,000	-	90,000	-	40,000
I	12"x 12"	80,000	-	100,000	-	45,000
Production weeks/yr		52.0	52.0	50.0	50.0	50.0

[10] See Ling (1988) for a comprehensive review of the function of the S&OP process.

The publication of appropriate data sets also fostered a 'give and take' environment at AO in which production schedulers, plant managers, and other field and headquarters personnel began to discuss the HPP system's data and results. This process again led to improved data accuracy and mutual participation and confidence in the plans generated by the integrated approach.

2.6 Summary

In this chapter, we have reviewed three case studies of firms that implemented hierarchical production and distribution planning systems in three different industries. There are valuable lessons to be learned from each of these implementations. In particular, each of the three HPP systems shared the common attribute that they provided hierarchical planning capabilities covering network-wide manufacturing and distribution operations. Therefore, each of these systems played a central role in both the long term and short term business planning activities of the three firms. Had these implementations not provided an integrated production and distribution perspective for the firms' entire networks, these HPP's could not have played such a 'core' decision support planning role.

Each of these three HPP systems made significant contributions towards improving the coordination and quality of the tactical planning activities and the operational execution of the three firms. We observed that all three firms reported significant reductions in annual operating costs and substantial improvements in customer service levels. These three HPP system projects also shared the attribute that the original implementation stimulated a process of ongoing improvements and enhancements to existing systems beyond the scope of the original HPP project. As the three case study firms began to understand the power of, and reap the benefits from their respective HPP systems, this caused the firms to push further to improve their HPP decision support capabilities. Again this represents a common evolutionary path found in many HPP projects.

In these case studies, we also observed three different approaches to building HPP systems. In one case, the Hax-Meal tire manufacturer case, the firm essentially constructed a brand new system with all new system components or modules. In the Agrico Chemical case, the firm developed one multi-purpose tactical and operational optimization modeling system and integrated existing decision support components around this optimization system. Finally, in the case of American Olean, the firm developed two new optimization models and several forecasting modules which it integrated with an existing inventory management system and several other components. In summary, there is no one right way to build a network-wide HPP planning and scheduling system. Rather, based on the availability and quality of existing decision support tools at a firm, the HPP designer must decide whether to build a complete system (as in the Hax-Meal approach), or whether to weave new planning components and existing components into an integrated HPP system.

Chapter 3

Production Planning And Scheduling

In this chapter, we focus primarily on hierarchical production planning and scheduling at the plant level. The case studies and discussion in Chapter 2 considered hierarchical planning from the perspective of managing operations across the entire manufacturing and distribution network of a firm. This perspective included evaluating, in an integrated fashion, manufacturing costs, distribution costs and capacities in the decision-making process of planning and managing a network's infrastructure and assets. Chapter 3 starts from the point where basic decisions about planning and utilizing the network have been developed based on an integrated manufacturing and distribution approach, and the firm's attention now turns more towards production planning. For example, in this chapter, we will assume that in the case of a network consisting of multiple manufacturing and distribution facilities, the firm has already established planned supply or sourcing patterns (e.g., which plants supply which products to particular distribution centers and demand regions), as well as the basic mission and responsibility of each facility. Methodologies such as those presented in Chapter 2 will have assisted in reaching this point.

The first section of Chapter 3 offers an overview of the wide range of hierarchical techniques used in production planning and scheduling. These techniques provide decision support from the strategic level down to the operational level with particular emphasis at the tactical and operational levels. After this overview, we next present selected illustrative methodologies for both annual (*i.e.*, tactical) and short run operational planning and scheduling. At the annual planning level, we consider the linkages required between the single plant and the overall network to assure a coordinated, feasible approach. At the operational level, we review several illustrative hierarchical scheduling approaches. Interspersed throughout this presentation we also consider methods for assuring the appropriate linkages, both top-down and bottom-up, to assure feasibility between schedules developed at the product family and end item levels respectively.

We conclude this chapter on production planning and scheduling with two different hierarchical perspectives. First, we present a model for simultaneously scheduling production at two plants on a multi-echelon manufacturing network. The hierarchical dimension in this case consists of explicitly linking the schedules of an intermediate (or first echelon) plant and a finishing (or second echelon) plant. We next examine a second classic hierarchical production scheduling methodology, namely, manufacturing requirements planning (MRP).

3.1 Overview Of Hierarchical Production Planning For Plants

This section provides a brief overview of the wide array of diverse applications of hierarchical production planning at plants. Although by no means comprehensive, the illustrations that follow offer insight into how pervasive the concept of hierarchical planning is in the field of production planning. In this section, we will qualitatively describe various types of hierarchical approaches, while later sections of this chapter will explore selected hierarchical modeling formulations and methods.

The rationale for using a hierarchical production planning and scheduling approach at the plant level is no different than the rationale for using a hierarchical approach to manage network-wide decision-making. A hierarchical methodology facilitates simplifying an overall problem into a series of smaller, more manageable sub-problems. At the same time, a hierarchical approach also offers a methodology for addressing the uncertainty and risk associated with production planning and scheduling. For example, it allows a firm to make aggregate decisions about overall plant level operations for the next 12 to 18 months, while postponing specific item level scheduling decisions until shortly before production actually occurs (e.g., a few days or even hours in advance).

General principles of hierarchical planning also apply to plant level production scheduling. For example, the more detailed types of scheduling that occur at a plant level (e.g., end item scheduling) typically must be updated quite frequently (e.g., weekly, or even daily). In contrast, more aggregate types of plant scheduling (e.g., product family scheduling) often require less frequent updating. The length of planning horizon considered in plant scheduling also depends upon the level of gradation and detail of the production planning. Plant planning horizons of over a year are common at the aggregate or upper planning levels, while planning horizons for detailed, disaggregated scheduling often extend only a relatively short time into the future (e.g., a few days to a few months). Concepts such as "frozen" scheduling periods also come into play at the plant level in the short term. While higher level plant planning may not even employ the concept of frozen periods or time buckets, short term schedules typically have many short time buckets (e.g., weeks, days or even hours) which, once scheduled and committed, cannot be changed in the very near term, (*i.e.*, they are frozen).

Table 3.1 presents a list of some of the more common hierarchical production planning applications found at the plant level. As previously noted, although by no means comprehensive, the applications described in Table 3.1 serve to convey the diversity of uses of hierarchical planning in plant production scheduling activities.

Table 3.1 Illustrative Hierarchical Plant Production Planning Applications

APPLICATION TYPE	PURPOSE	TYPICAL PLANNING HORIZON
Aggregate Workforce Planning and Detailed Scheduling	• Plan aggregate workforce requirements and operating levels • Plan machine purchases and machine utilization levels • Plan detailed product schedules and changeover schedules • Plan overtime and short term labor schedules	• 12 to 18 months • 12 to 18 months • a few weeks to 6 months • a few weeks to 6 months
Aggregate Capacity Planning And Product Family Scheduling	• Create long term capacity utilization and product mix plans at the product family level • Evaluate impact of product family changeovers and short term schedules on feasibility of long term plan • Evaluate costs of alternative changeover and production mix plans	• 12 to 18 months • 6 to 12 months in short term scheduling buckets (e.g. weekly buckets)
Job Shop Design, Capacity and Scheduling	• Determine appropriate capacity of a job shop including the number and mix of machines • Schedule work on machines - typical objective is to minimize maximum completion time of all jobs - scheduling often includes sequencing	• 12 to 18 months • a few weeks to 6 months
Multi-Stage/Multi-Echelon Plant Scheduling	• Schedule "linked" WIP and finished product plants in an integrated approach - short term scheduling can be done in one simultaneous model	• a few weeks to 3 to 6 months
Material Requirements Planning [1]	• Determine production, purchase and materials requirements at the sub-independent end item (*i.e.*, dependent) level – based on projected independent end item requirements	• a few weeks to 12 to 18 months
Short Run Family And End Item Scheduling	• Create short run product family schedules (e.g., in weekly buckets) • Develop end item schedule based on family schedule - evaluate if any infeasibilities occur at the item level which are not visible at the family level • Sequence items within families	• 3 to 12 months in short term scheduling buckets (e.g., weeks) • 1 to 3 months • a few days to 3 months
Batch And Load Scheduling In Flexible Manufacturing Systems	• Determine how to best create batches of like part types to produce together • Determine loading (*i.e.*, production) schedules for batches created - can be done sequentially or simultaneously	• 1 to 6 months
Batch And Load Scheduling, And Routing And Materials Handling Path Control In Flexible Manufacturing Systems	• Determine part types and batch volumes • Plan shift or daily production levels • Determine actual routes for parts - typical objective is to minimize material handling requirements • Execute routing, sequencing and material handling path control	• 1 to 6 months • 1 week to 4 to 8 weeks • very short term, a week or two at most • a few days
Multi-item, Single Machine Scheduling	• Plan machine capacity • Schedule items and sequence	• 1 to 6 months

[1] this refers to pure materials requirements planning, not full-scale MRPII

These applications are "loosely" ordered in terms of the typical length of the planning horizon spanned by the application. We use the term "loosely" because one can find a broad set of planning horizons associated with each of these applications in practice. Nevertheless, the first three applications (aggregate workforce planning, aggregate capacity planning and job shop design and capacity planning) generally encompass the longest planning horizons. For example, the planning hierarchy associated with aggregate workforce planning through detailed scheduling can range from determining labor force plans for up to 18 months into the future to scheduling labor for shifts over the very near term (e.g., days to a couple weeks.) Similarly, aggregate workforce planning and scheduling implementations also generate production plans at the product family level for 12 to 18 months, while also providing detailed production schedules for the near term at the line item level.

Aggregate capacity planning and product family scheduling modeling systems, as well as job shop design, capacity and scheduling systems cover a similar length tactical planning horizon (*i.e.*, 12 to 18 months) while also providing detailed scheduling for the short run in relatively small time buckets (e.g., weekly time buckets). Aggregate capacity planning models are often found in process industries because firms in these industries typically focus on capacity first, and then schedule materials and labor (e.g., see Taylor et. al., 1981). The ceramic tile case study presented in section 2.4 represents an example of such a planning system.

At the tactical level, job shop hierarchical planning systems focus on determining the number, mix and total capacity of machines required to support operations for up to 12 to 18 months. In contrast, at the shorter run operational level, job shop scheduling systems typically develop work schedules for machines in time buckets such as shifts, days and weeks. Decisions scheduled at this level often include not just machine usage level, but also job sequences through the shop and so on.

The remaining hierarchical planning applications illustrated in Table 3.1 are generally used to assist production planning and scheduling from 6 months or so in (*i.e.*, shorter planning horizons). Software used to support these applications may in some cases have planning horizons that span 12 or more months. Nevertheless, the focus of planning and scheduling typically remains closer in. For example, multi-echelon, synchronized plant production scheduling models typically do not establish schedules beyond three to six months out. Besides the fact that these modeling systems can present a major computational challenge if the number of model periods becomes too large, the focus on simultaneously scheduling multiple "linked" plants generally remains for the next few weeks and months. We classify synchronized plant scheduling models as "hierarchical" planning systems for slightly different reasons than the other applications illustrated in Table 3.1. As an example later in this chapter illustrates, these short run, operational models simultaneously schedule a hierarchy of plants; namely, an intermediate echelon work-in-process (WIP) plant and a second echelon finished goods inventory (FGI) plant. Although these models themselves can be one submodel of a larger overall hierarchical system, there exists a hierarchical dimension within just this single model itself (*i.e.*, the multiple echelons).

Traditional MRP software systems often do have planning horizons that extend out 12 months or more, and typically these systems view their planning horizons in weekly buckets. Despite having the capability to address longer horizons, MRP[1] systems at the plant level generally focus on near term, operational scheduling (*i.e.*, 6 months or less). MRP systems, as is well know, are hierarchical in the sense that driven by projected time-phased requirements for independent end items, these systems derive dependent materials requirements through defined hierarchical bill of materials relationships between end items and their component parts. In contrast to MRP, the remaining applications shown in Table 3.1 focus on scheduling product families and end items, as well as the resource requirements, and in some cases the routings associated with family and item schedules. We will review techniques and issues to consider in short run family and end item scheduling later in this chapter.

This completes our brief review of selected hierarchical production planning and scheduling applications at the plant level. As noted, the remaining sections of this chapter present selected models and techniques applicable to plant production scheduling.

3.2 Linking Tactical Network Planning Methods And Single Plant Productions Plans

As reviewed in Chapters 1 and 2, the integrated tactical network-wide production and distribution planning process generally results in a production plan at the product family level for each individual plant on a firm's network. A key next step in this planning process consists of evaluating this plan in detail at the individual plant level. Figure 3.1 indicates where the linking of network and single plant production plans occurs in the overall hierarchical planning process originally depicted in Figure 1.7. This linking process represents an extremely important step for several reasons.

As noted in Chapter 1, the annual (tactical) network-wide planning process that develops the product family production plans for each plant generally creates product family plans either in single period annual time buckets or in four quarterly time buckets. In contrast, individual plants typically schedule product family production in weekly time buckets or other similar relatively short time periods. The network-wide models that generate the initial plant product family plans in quarterly or single time buckets have not considered the impact of (weekly) product family changeovers at the individual plant level. Thus, it is important to determine whether these initial product family plans remain feasible when evaluated at the plant level from a weekly time bucket scheduling perspective. Individual plants may also have minimum run quantity constraints on some or all

[1] Again, we are referring here to traditional MRP plant scheduling systems and not to the full scale MRPII systems that today address a much broader set of planning dimensions than just plant scheduling.

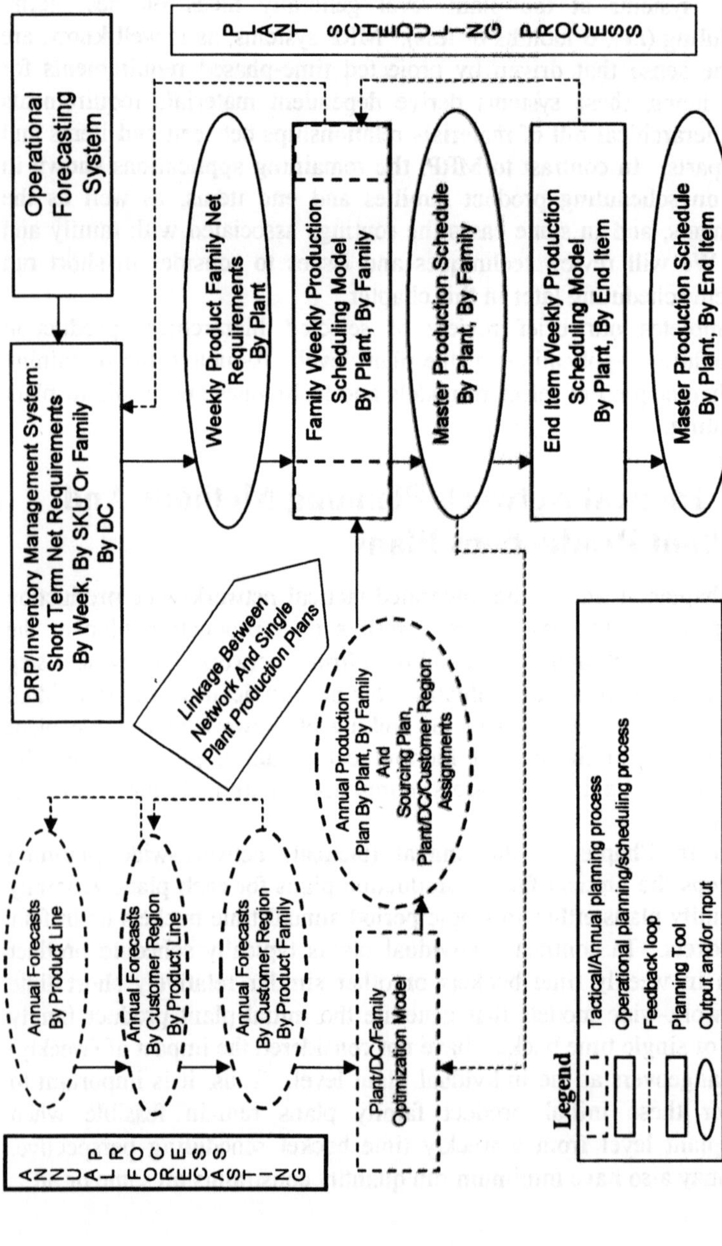

Figure 3.1 Integrated Production And Distribution Planning System (Tactical/Annual Planning And Scheduling)

product families. For example, an individual plant may have a restriction that it will not produce less than three consecutive weeks (periods) of certain product families. Planners can usually at least partially address such operating restrictions by including minimum run quantities in the annual network-wide planning models. However, given that such models may have quarterly or single annual time buckets, it is not possible to completely assure that network-wide generated individual plant plans do not implicitly violate the "consecutive" week restriction. Evaluating the feasibility of consecutive minimum run quantity type restrictions at the plant level is in practice usually a relatively minor issue compared to that of evaluating the impact of weekly changeovers. Nevertheless, it serves to illustrate the types of infeasibilities that can surface at the plant level, but which do not appear at the network-wide hierarchical modeling step. In some industries, product family changeovers are sequence dependent upon such factors as product color (e.g., the sequence may be light colors to darker colors). In such cases, evaluating the feasibility of changeovers at the plant level can take on even additional importance as this added dimension may increase the possibility that the initial network-wide production plan contains undetected infeasibilities.

There exist numerous methods for taking the initial product family production plan generated by the integrated network-wide modeling process and evaluating it at the plant level on a plant by plant basis. For illustrative purposes, we present one such approach on the next page. This illustration assumes the hierarchical planning process described in Figure 3.1.

Table 3.2 illustrates steps 1 and 2 of the process for evaluating individual plant product family production plans. As the table depicts, these represent very basic, mechanical steps. Steps 3 and 4 where the planner must make modifications for inventory builds and/or reductions, and then reconcile the weekly plant product family demand schedule totals with the network-wide generated production plan totals can require some minor data manipulation. Nevertheless, this entire process (steps 1 - 4) is quite straightforward and a planner can implement this technique in practice quite easily. To execute steps 5 through 7 requires that a planner run the data created in steps 1 to 4 through a plant weekly product family scheduling model. We refer the reader to the short run scheduling model previously presented in section 2.4, as well as the model which follows shortly in section 3.3, as illustrations of the types of models that can facilitate these steps. It is important to note, however, that one can perform steps 5 to 7 using non-optimization based spreadsheet models, simulation models, etc. Even if a firm employs a very manual approach to developing weekly plant product family production schedules, the firm can and should still evaluate the feasibility of production plans developed at an aggregate, annual network-wide level at the individual, weekly plant level.

Illustrative Approach For Evaluating Individual Plant Product Family Production Plans

- Step 1: Create a weekly (period) seasonal index of demand (*i.e.*, production requirements) by product family for a plant. (Note: this can be based on either total network demand for a product family, or more typically, demand at the individual plant, whichever is more appropriate.)
- Step 2: Disaggregate the annual product family production plan generated by the network optimization model for a plant into weekly (period) production plans (*i.e.*, requirements). Use the index created in step 1 to disaggregate the annual plan for each product family.
- Step 3: Adjust the plant weekly product family production requirements created in step 2 for any appropriate factors not accounted for by the demand index. For example, adjustments would have to be made for any planned inventory builds or reductions imbedded in the annual product family plans developed using the network-wide optimization model. Similarly, modifications must be made for planned inventory builds or decreases associated with new product introductions or product discontinuations.
- Step 4: Confirm that the weekly (period) production requirements generated for each product family sum to the quarterly and annual production totals assigned to the plant by the network optimization model.
- Step 5: Input the weekly (period) product family production requirements into the plant product family weekly production scheduling model and run the model.
- Step 6: Evaluate the feasibility of the weekly (period) production schedule developed by the weekly scheduling model. If the weekly model can produce a satisfactory, feasible schedule then the annual product family production plan created by the network optimization model is a feasible plan for the individual plant. Further, the product family production plan (*i.e.*, the feasible schedule) created by the weekly scheduling model represents the plant's annual production plan scheduled by week.
- Step 7: If the weekly model has produced a good product family production plan, then stop. If the weekly model cannot generate a feasible plan, or if it cannot generate an acceptable feasible plan; then the annual (tactical) planning process must continue. The first step will be to determine if some minor revisions to the plan can be developed at the individual plant level (e.g., within some defined tolerance) which will yield an acceptable plan. On the other hand, if significant infeasibilities or other major issues are identified by the evaluation process in steps 1 - 6, then planners may have to revise the annual network-wide plan to address the infeasibilities surfaced at the individual plant "weekly scheduling" level. For example, additional constraints may have to be appended to the network-wide optimization model so that it can generate a product family production plan for the individual plant which is feasible. This would represent a feedback loop in a hierarchical planning system and Figure 3.1 shows this loop from the plant level back to the network-wide optimization based planning level. Note that this type of feedback and revision to the network-wide plan can have implications for one, several or all of the plants on a network (*i.e.*, revisions to the production plan and product mix of one plant may necessitate changes at several other plants on the network also).

Table 3.2 Evaluating Individual Plant Product Family Production Plans (Steps 1 and 2)

	Annual Product Family Production Plans Developed By Network-wide Optimization Model			Weekly Cumulative Demand Index For Regions Served By The Plant - (step 1)			Initial Dissaggregation Of Annual Product Family Plans Into Weekly Product Family Production Plans (I.e., Demand Requirements) - (step 2)	
Quarter	Family A (000 units)	Family B (000 units)	Wk	Family A	Family B	Wk	Family A (000 units)	Family B (000 units)
			1	0.90	1.05	1	90,000	210,000
			2	0.95	0.95	2	95,000	190,000
			3	1.10	1.05	3	110,000	210,000
			4	1.20	1.15	4	120,000	230,000
			5	0.95	0.95	5	95,000	190,000
1	1,300,000	2,600,000	6	0.90	0.90	6	90,000	180,000
			7	1.05	0.95	7	105,000	190,000
			8	1.05	1.00	8	105,000	200,000
			9	1.00	1.00	9	100,000	200,000
			10	0.95	1.05	10	95,000	210,000
			11	0.85	0.90	11	85,000	180,000
			12	1.00	0.95	12	100,000	190,000
			13	1.10	1.10	13	110,000	220,000
Sub-total				13.00	13.00		1,300,000	2,600,000
2	1,500,000	3,000,000	• • •	• • •	• • •	• • •	• • •	• • •
3	1,400,000	2,800,000	• • •	• • •	• • •	• • •	• • •	• • •
4	1,700,000	2,600,000	• • •	• • •	• • •	• • •	• • •	• • •

3.3 Operational And Short Run Plant Scheduling Methods

The number of methods and algorithms used to perform operational, short run planning and scheduling at plants is almost countless. Even in relatively similar production environments (e.g., all manufacturers in a specific industry), one can usually find innumerable different scheduling methodologies in use at different firms, and oftentimes even at a single firm. To illustrate plant scheduling methods, this section presents several operational scheduling models. These models will provide only a glimpse of the myriad types of decision support systems that exist at this planning level. Nevertheless, they serve to illustrate some of the key types of decisions that occur in plant scheduling.

Operational, short run plant scheduling activities, similar to network planning activities, have a broad hierarchical dimension just within their own domain. For example, these plant scheduling activities span from "relatively" higher level planning activities such as scheduling product family production and weekly changeovers between product families at a line and/or plant level, to the most detailed shop floor activities at a particular machine. The planning horizon of these operational activities can extend from as long as eighteen months into the future, to as little as a few hours or days ahead.[2] Figure 3.2, a representation of the generic hierarchical process presented in Figure 1.7, illustrates where these plant scheduling activities fit into the planning and scheduling process.

3.3.1 Hierarchical Plant Operational Planning And Scheduling Methods

Section 2.2 reviewed one of the classic hierarchical plant planning and scheduling systems, namely, the implementation at a tire manufacturer credited to Hax and Meal (1975). Our original discussion presented this as a full scale hierarchical system equipped to guide planning and scheduling over an entire network. However, the lower level components of this system represent a *plant* planning and scheduling system. Figure 3.3 provides another overview of the Hax and Meal planning and scheduling system. Note that from the seasonal planning module on, this system is an individual plant production planning and scheduling system. The reader is referred to section 2.2 for additional discussion of this system.

---

[2] It should be noted, however, that when an operational, short run plant scheduling system maintains a planning horizon of up to eighteen months, the focus for operational purposes is usually just one to two quarters ahead at the most. Planning exercises that extend twelve to eighteen months into the future generally fall more into the tactical (annual) planning category rather than operational category. As Figure 3.2 illustrates, some operational planning methodologies (e.g., family weekly production scheduling models) can support both operational and tactical production planning.

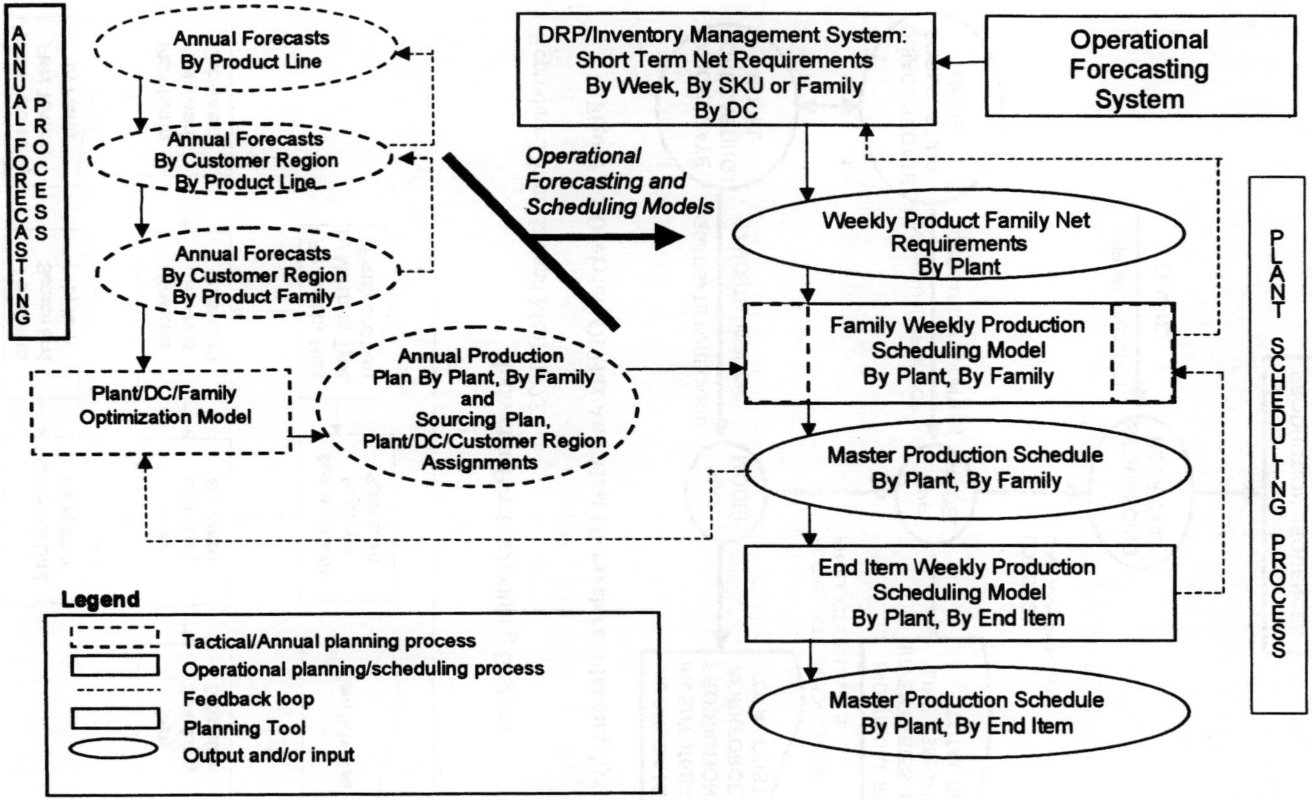

Figure 3.2 Generic Integrated Production And Distribution Planning System (Tactical/Annual Planning And Scheduling)

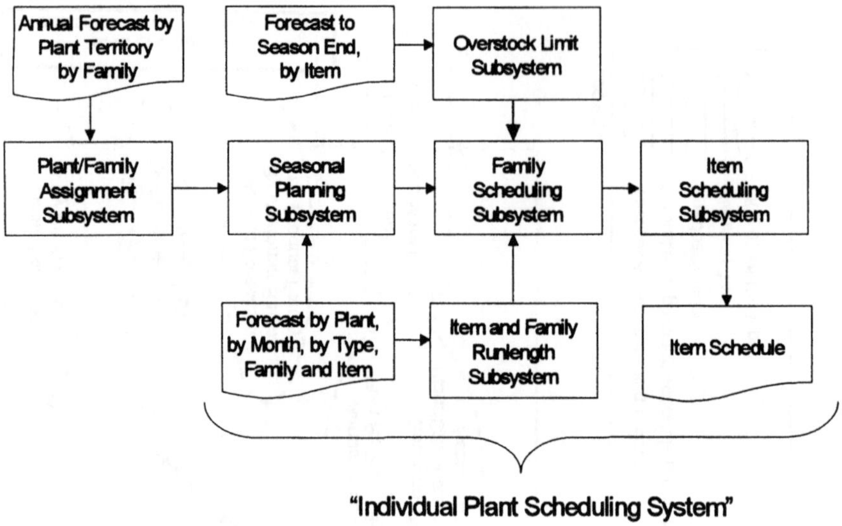

- Reproduced from Hax and Meal, 1975.

Figure 3.3 Overview Of Hax And Meal Hierarchical Planning System

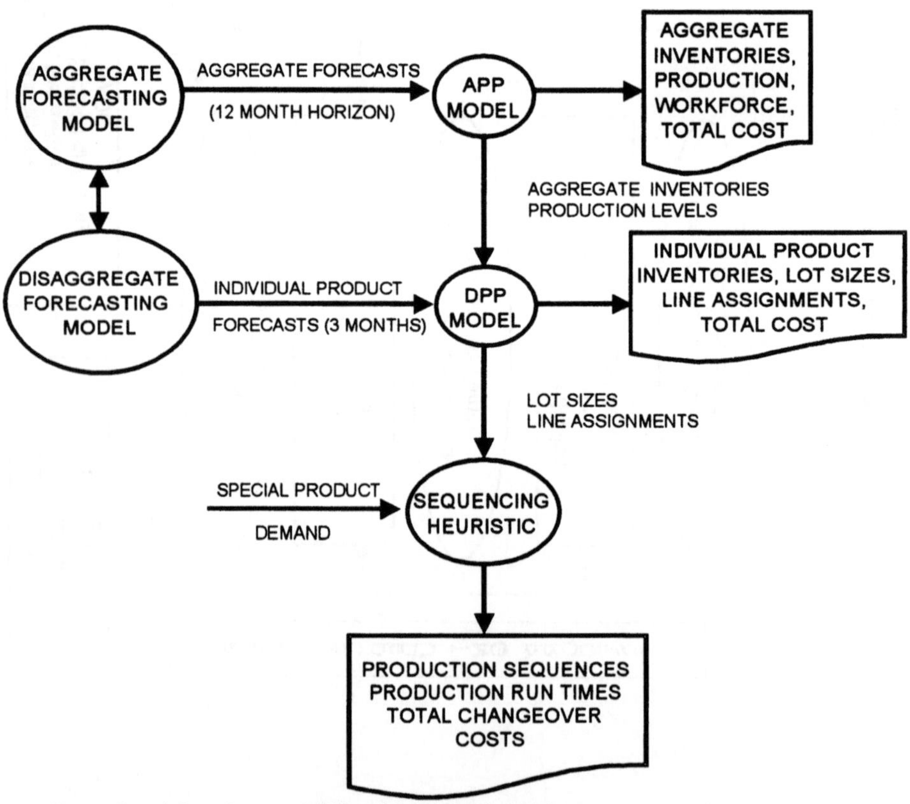

- Reproduced from Leong, Oliff and Markland, 1989.

Figure 3.4 Plant HPP Model Used By Owens-Corning Fiberglass

Another well known hierarchical production planning and scheduling system was the one implemented by Owens-Corning Fiberglas during the 1980's. Numerous articles have described aspects of this implementation including those of Burch, Oliff and Sumichrast (1987), Earl and Sumichrast (1987), and Leong, Oliff and Markland (1989). Figure 3.4 illustrates the hierarchical planning and scheduling system constructed by Owen-Corning Fiberglas to schedule production at a plant that manufactured fiberglass mats in a batch production environment.

Briefly, this plant scheduling system had three primary planning and scheduling levels as illustrated in Figure 3.4. At the highest planning level, Owens-Corning projected aggregate workforce and overtime requirements, as well as inventory requirements and levels for a twelve month planning horizon. The firm employed an "aggregate production planning" model based on an algorithm with production switching rules to determine its twelve month long term operations plan for the plant (see Leong, Oliff and Markland, 1989). An aggregate forecasting model which projected demand at both the product group level (for lower volume items) and at the individual item level (for key high volume products) drove the aggregate planning model.

At the next lower planning level, Owens-Corning employed a "disaggregate production planning" model to generate production line assignments and lot sizes for individual products, as well as detailed inventory schedules by product. A disaggregate forecasting model providing three month projections in monthly time buckets drove the disaggregated production planning model. Interestingly, Owens-Corning utilized a goal programming based approach at this planning level. Leong et. al. (1989) explained that the

> "goal programming model... allows deviations from the aggregate decisions (goals), namely aggregate production capacities and inventory levels determined from the aggregate production planning model... to address the disaggregation problem.... The approach also allows different costs/weights to be assigned to deviational variables to reflect the relative importance of the goals and to provide explicit feedback regarding aggregate production planning objectives."

At the next and lowest operational scheduling level, Owens-Corning employed a job sequencing model to determine the day-to-day production line schedules and changeovers. This model minimized changeover costs subject to product demand requirements.

The Owens-Corning HPP model is representative of the hierarchical planning and scheduling activities commonly found at the plant level. Initially, some type of aggregate planning model provides a projection of the overall level of work required of a plant for a long term planning horizon (e.g., twelve months). This projection will help size such variables as the overall workforce, machine and/or materials requirements of a plant depending upon what the key drivers of a particular plant and industry may be. After an aggregate planning model develops these higher level plant requirements, then shorter run decision support models are used to develop detailed schedules for plant operations. Importantly, these lower level models also provide feedback to the higher level models. As noted in previous chapters, this feedback process becomes particularly critical when a lower level model uncovers infeasibilities not identifiable at a higher level.

Flexible Manufacturing Systems

Flexible manufacturing systems (FMS) characteristically have a level of complexity that necessitates sophisticated planning and scheduling techniques such as hierarchical production planning. FMS generally include numerically controlled workstations, a material handling system, and a (central) controller that coordinates the production of assorted part types. Numerous models and systems exist to plan and schedule the operations of a FMS (see e.g., Jaikumar and Wassenhove, 1989; Tang and Denardo, 1988A; and Tang and Denardo, 1988B). An example of a hierarchical production planning approach to schedule a FMS plant is the model presented by Sodhi, Askin and Sen (1994), and Figure 3.5 describes their approach.

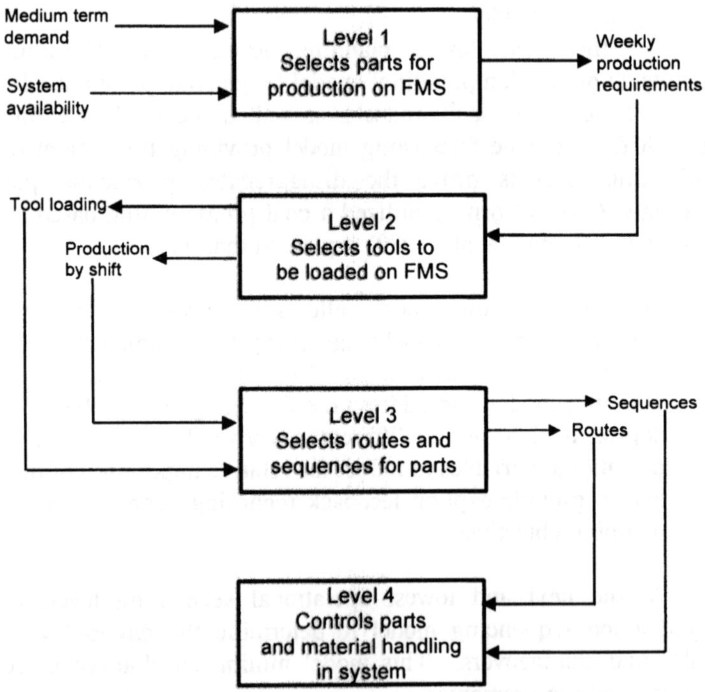

▪ Reproduced from Sodhi, Askin and Sen (1994).

Figure 3.5 A HPP Model For A FMS Plant

The top level of this HPP typically schedules production volumes over a several month, medium term planning horizon in individual weekly time buckets.[3]

---

[3] These time buckets can actually range from one week to one month.

The primary planning output at this level is the determination of the part types and the quantities to be produced by a FMS over a medium term planning horizon. The optimization model utilized at this level minimizes system operating costs, shortage costs and inventory carrying costs subject to standard constraints such as machine capacity and material handling resource availability.

The second level of this HPP schedules the daily shift production of the FMS for the next week (*i.e.*, the first period of the level 1 model). At this level, the scheduling effort focuses on assigning tools to machines (*i.e.*, the system configuration) to facilitate the planned part type production. An algorithm which seeks to minimize inventory holding and shortage costs, while recognizing the tool change requirements of alternative plans, drives the schedule generated at this level. Importantly, to enhance coordination across planning levels, Sodhi, Askin and Sen (1994) formulate the cost objective of the level 2 scheduling model as a problem to minimize deviations from the cost minimizing plan previously generated for period 1 by the level 1 planning model.

As Figure 3.5 illustrates, the third level scheduling process determines the actual routes (through the machines) for the individual part types to follow. The routes determined at the third level allow the production volumes planned at the previous (second) level to occur. Before this routing assignment process takes place, however, a check occurs to evaluate whether there exists any unused capacity available in the coming period. Specifically, if there are any "empty" tool slots not scheduled for use, Sodhi, Askin and Sen employ a linear network type algorithm to "add" tools to any empty "magazines". The evaluation of whether to add tools to empty slots is performed by evaluating the preliminary tool assignments first made at level 2 prior to routing considerations. The algorithm used to add tools essentially projects the potential bottleneck or shared machine capacity volumes implied by the initial tool assignments generated by the level 2 model.

Once tools have been added to any potential empty magazines, a linear programming model evaluates the feasible process routes for each part type scheduled for production and establishes the routings to be used in the next production period. An objective function that minimizes either total transport distances or total material handling costs drives this model.

Finally, shop floor type decision-making and control occurs at level 4, the lowest level of the HPP system. Here dispatching rules and other materials handling control rules are implemented to execute the detailed production plan created at level 3. Thus, this HPP for a FMS plant addresses a full gamut of planning and scheduling activities from a medium term operational planning perspective (e.g., several months to a year) down to detailed shop floor control activities.

Summary

In this sub-section on hierarchical plant operational planning methods, we have briefly reviewed three approaches used in the tire, fiberglass and flexible manufacturing industries. As one would expect, the actual scheduling models employed in each application vary considerably. However, they do share the common approach of first planning operations over a longer planning horizon at a more aggregated product planning level (e.g., product families). Each HPP then moves from an aggregated planning level down to a detailed, short run scheduling level in a series of several steps. We conclude this section with Figure 3.6 which provides yet another schematic of an HPP implementation at a plant. Bowers and Jarvis (1992)[4] described the implementation represented in this figure as applicable to industries such as steel, textiles and paper. Again we observe a planning system that progresses from the annual to the weekly to the daily levels in its scheduling activities.

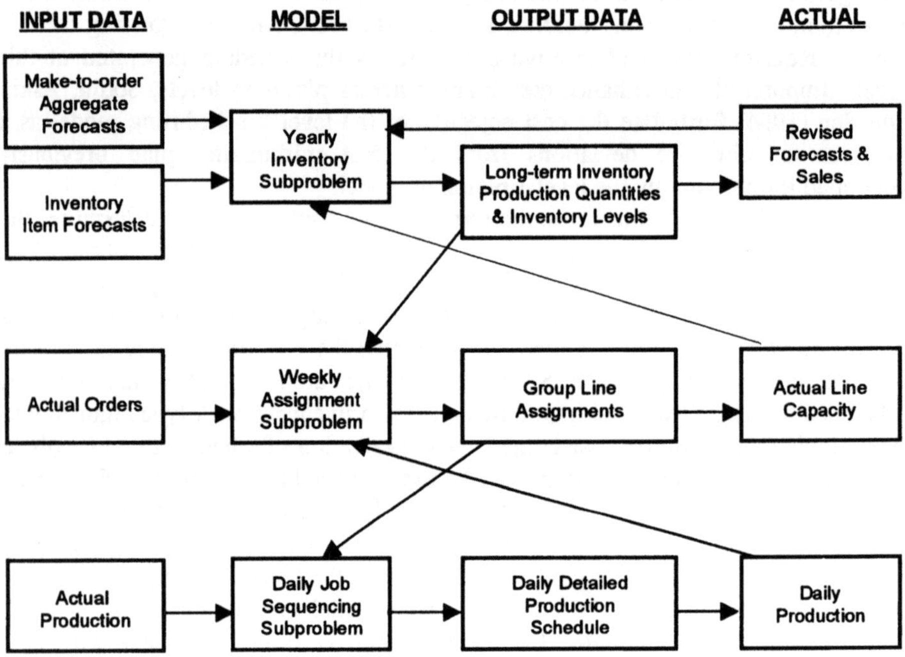

• Reproduced from Bowers and Jarvis, 1992.

Figure 3.6 Overview Of Bowers And Jarvis Hierarchical Planning System

3.3.2 A Product Family Plant Scheduling Optimization Model

Section 3.2 discussed linking network and single plant production plans, and section 3.3.1 reviewed general hierarchical frameworks for planning and scheduling individual plants. In this section, we review in more detail a single plant optimization based scheduling model. This product family level model can function both as a decision support tool to evaluate network-wide production plans at the individual plant level, and as a plant product family scheduling model. A review of this model will illustrate the types of decisions evaluated in higher level, aggregate plant production planning models. This review will also provide perspective for section 3.3.3 which describes an illustrative feedback loop between

[4] The authors did not reveal their specific application.

an end item scheduling model and a family scheduling model such as the one depicted in this section.

This product family scheduling model has particular applicability to industries such as the process industries where production line start-up and shutdown costs are often quite high. The substantial costs of starting up and shutting down production lines poses a difficult dilemma for a process industry manufacturer when a plant's capacity significantly exceeds its known and forecasted demands over a tactical (e.g., annual) planning horizon. In this situation, the manufacturer must either plan to temporarily shut production lines down during the planning horizon or idle the lines without formally shutting them down. Naturally, a firm can also choose a mix of the strategies. The key difference between these alternatives is that by idling a line rather than shutting it down, one avoids the substantial shutdown and start-up costs. This temporary line shutdown versus idle decision represents another type of decision that a firm frequently addresses at the plant level rather than at the network-wide tactical planning level. A decision to shut an entire plant (or even one production line of a plant) down over an entire planning horizon typically occurs at the network-wide planning level. However, decisions about idling vs. shutdown on one or more lines, for a part of a planning horizon, frequently fall to the individual plant planning level.

Background

The trade-offs between shutdown and idle strategies can be complex. For example, by temporarily idling a production line, a process manufacturer can save on shutdown and start-up costs, perhaps reduce the loss of goodwill associated with temporary furloughs and rehiring, and so on. Additionally, by keeping a line idle, but ready to produce, a firm can maintain a higher degree of manufacturing flexibility in cases in which, for example, two production lines can make the same product and shutting one line down limits the capability to produce certain products to one line. However, below some level of demand for plant capacity, the cost inefficiencies of keeping a line available, but idle, more than outweigh any relative advantages of maintaining a line's immediate availability.

As previously noted in section 3.2, for planning purposes at the plant level, the annual (tactical) production plan must be disaggregated into weekly production requirements. It is at this point that if the production levels allocated to a plant in the annual network-wide planning model do not fully utilize the individual plant's capacity over the planning horizon, then a decision is made whether to fully shut one or more lines down for an extended period of time or to idle temporarily one or more production lines. This represents the decision point in the planning process where the following model can assist. Finally, it is important to note that the following product family level model can serve as a standard scheduling model when capacity does not significantly exceed forecast demand (*i.e.*, when the temporary shutdown vs idle decision disappears).[5]

[5] In this case, model (3.1) – (3.13) essentially reduces to model (2.29) – (2.33).

Problem Statement

The following product family scheduling model minimizes the total of variable manufacturing costs, production line changeover/setup costs, inventory carrying costs, idling costs and shutdown costs. The cost of restarting a production line after it has been shutdown is included in the cost of a line shutdown to recognize that a shutdown decision also implies future start-up costs. This also prevents the model from generating solutions that shut down production lines near the end of planning horizons without considering that a start-up cost will be required in a future period beyond the current planning horizon. Additionally, whatever appropriate one time costs associated with the shutdown and start-up process should be included. For example, in a process industry such as the ceramic tile industry, the cost to repair a kiln should be included if a shutdown generally causes some damage to the kiln. For purposes of the formulation, it is assumed that production rates remain constant during the scheduling period. Additionally, in the manufacturing environments that this model was originally developed to support (e.g., a ceramic tile plant), a major changeover under normal conditions does not reduce the output of a production line, and therefore, changeover time can be ignored. Finally, backlogging of demand is not allowed in the following model. Table 3.3 displays the notation used to formulate the model.

Table 3.3 Notation For The Plant Product Family Scheduling Model

N	is the number of products.
L	is the number of lines.
T	is the number of periods (in weeks).
i	is the index on the product ($i = 0, 1, \ldots, N$). Product zero is not a real product. If a production line j is assigned to product zero during period t, this means that j is shutdown during period t.
j	is the index on the production lines or processors ($j = 1, \ldots, L$).
t	is the index on the production period ($t = 1, \ldots, T$).
L_i	is the set of lines that can produce product i.
M_j	is the set of products that can be produced on line j.
c_{ij}	is the unit variable manufacturing cost to produce product i on line j.
r_{ij}	is the cost per period of keeping line j temporarily idle while it is setup to produce i.
h_i	is the unit carrying cost per period of product i.
s_{oj}	is the fixed cost per period to keep a production line j shutdown, beginning with the first period in which the production line is shutdown.
q_{oj}	is the fixed cost to shut down production line j during the first period of a line shutdown. The fixed cost to restart production line j at some future point is also included in q_{oj}. Also, note that the value of q_{oj} must exclude s_{oj} to prevent double counting s_{oj} during the first period.
q_{ij}	is the fixed cost to changeover to product i on line j, for $i = 1, \ldots, N$.
p_{ij}	is the fixed production quantity per period of product i on line j.
d_{it}	is the demand of product i in period t.
I_{io}	is the starting inventory of product i.

Table 3.3 Notation For The Plant Product Family Scheduling Model (continued)

$IMAX_{it}$	is the maximum inventory level allowed (or upper inventory limit) of product i at the end of period t.
$IMIN_{it}$	is the minimum inventory level (or lower inventory target) planned for product i at the end of period t.
A_j	is the maximum number of consecutive periods that a production line j can be temporarily idle.
B_j	is the maximum number of periods of temporary idling allowed over the entire planning horizon on line j.
B	is the maximum number of periods in total of temporary idling allowed over the entire planning horizon over all lines $j \in L$.

As noted, in this formulation we adopt the convention that if a line j produces product 0 during a period t, this indicates that the line is shutdown during that period. The decision variables of this formulation are as follows:

$$W_{ijt} = \begin{cases} 1, & \text{if product } i \text{ is produced on line } j \text{ in period } t; \\ 0, & \text{otherwise} \end{cases}$$

$$Y_{ijt} = \begin{cases} 1, & \text{if line } j \text{ is setup to produce product } i \text{ in period } t; \\ 0, & \text{otherwise} \end{cases}$$

$$Z_{ijt} = \begin{cases} 1, & \text{if production on line } j \text{ changes to product } i \text{ in period } t; \\ 0, & \text{otherwise} \end{cases}$$

I_{it} = the end of period inventory of product i in period t.

The problem formulation is as follows:

$$\text{Minimize: } \sum_{t=1}^{T}\sum_{i=1}^{N}\sum_{j\in L_i}\left(c_{ij}p_{ij}W_{ijt} + r_{ij}\left(Y_{ijt} - W_{ijt}\right) + q_{ij}Z_{ijt}\right) + \sum_{t=1}^{T}\sum_{i=1}^{N} h_i I_{it} + \sum_{t=1}^{T}\sum_{j=1}^{L}\left(q_{oj}Z_{ojt} + s_{oj}Y_{ojt}\right) \tag{3.1}$$

Subject to:

$$I_{it} \geq IMIN_{it} \qquad i = 1, 2, \ldots, N, \qquad t = 1, 2, \ldots, T \tag{3.2}$$

$$I_{it} \leq IMAX_{it} \qquad i = 1, 2, \ldots, N, \qquad t = 1, 2, \ldots, T \tag{3.3}$$

$$\sum_{i\in M_j} Y_{ijt} + Y_{0jt} = 1 \qquad j = 1, 2, \ldots, L, \qquad t = 1, 2, \ldots, T \tag{3.4}$$

$$Z_{ijt} \geq Y_{ijt} - Y_{ijt-1}, \qquad j \in L_i \quad i = 0, 1, 2, \ldots, N, \qquad t = 1, 2, \ldots, T \tag{3.5}$$

$$W_{ijt} \leq Y_{ijt} \qquad j \in L_i \quad i = 1, 2, \ldots, N, \qquad t = 1, 2, \ldots, T \tag{3.6}$$

$$\sum_{i \in M_j} \sum_{k=t-A_j+1}^{t=1} \left(Y_{ijk} - W_{ijk}\right) \leq A_j \quad t = A_j, \ldots, T-1, \; i \in M_j, \quad j = 1, 2, \ldots, L \tag{3.7}$$

$$\sum_{i \in M_j} \sum_{t=1}^{T} \left(Y_{ijt} - W_{ijt}\right) \leq B_j \qquad j = 1, 2, \ldots, L \tag{3.8}$$

$$\sum_{i=1}^{N} \sum_{j \in L_i} \sum_{t=1}^{T} \left(Y_{ijt} - W_{ijt}\right) \leq B \tag{3.9}$$

$$I_{it} = I_{i0} + \sum_{j \in L_i} \sum_{k=1}^{t} p_{ij} W_{ijk} - \sum_{k=1}^{t} d_{ik} \qquad i = 1, 2, \ldots, N, \qquad t = 0, 1, 2, \ldots, T \tag{3.10}$$

$$0 \leq W_{ijt} \leq 1, \textit{ integer} \quad j \in L_i, \qquad i = 1, 2, \ldots, N \qquad t = 1, 2, \ldots, T \tag{3.11}$$

$$0 \leq Y_{ijt} \leq 1, \textit{ integer} \quad j \in L_i, \qquad i = 0, 1, 2, \ldots, N, \qquad t = 1, 2, \ldots, T \tag{3.12}$$

$$Z_{ijt} \geq 0 \qquad j \in L_i, \qquad i = 0, 1, 2, \ldots, N, \qquad t = 1, 2, \ldots, T \tag{3.13}$$

Constraints (3.2) prevent the backlogging of demand and ensure that the minimum inventory requirements are satisfied. Constraints (3.3) preclude solutions that produce excess inventory to avoid the potentially high costs of shutting down a line. Whether this would occur naturally depends on the comparative costs associated with shutting down a production line, idling a production line, and producing and carrying excess inventory. Note that one can set the maximum allowable level of inventory of product i through the parameter $IMAX_{it}$. Constraints (3.4) and (3.12) assure that any production line is setup to produce exactly one product (including product 0) during each period. They constrain each line j during period t to one of two conditions: the line is producing a product or the line is shutdown and requires start-up procedures before it can resume production. Constraints (3.5) relate the changeover variable (Z_{ijt}) to the production setup variables in the current and previous periods (Y_{ijt} and $Y_{ij,t-1}$, respectively). A changeover to product i on line j occurs during period t, if and only if $Y_{ijt} = 1$ and $Y_{ij,t-1} = 0$. It is easily demonstrated that constraints (3.5) force Z_{ijt} to be one only if $Y_{ijt} = 1$ and $Y_{ij,t-1} = 0$; in all other cases Z_{ijt} must be zero. Therefore, Z_{ijt} need not explicitly be specified as a binary variable.

Constraints (3.6) assure that production takes place only when a production line j is setup to produce a product. However, when no production takes place on a line that is setup to produce a real product (*i.e.*, any product but product zero), then we say the line is temporarily idle. To understand what we mean by saying that a production line is temporarily idle, consider an illustrative example involving the manufacture of ceramic tile. While in operation, a ceramic tile kiln requires that unfired tiles move through it constantly to absorb the heat generated by the kiln. If a kiln is lit and there is no tile moving through the kiln absorbing the heat, a fire will likely result. Thus, if a manufacturer wishes to idle a production line without turning off the kiln (*i.e.*, without shutting the line down), it must push nonsaleable dummy or slug tiles through the kiln to absorb the heat and prevent a fire. Manufacturers will do this in cases where they wish to stop or slow production temporarily without having to incur the costs and/or allocate the resources to turn off a kiln and then refire the kiln at some future time. To maintain a production line in this state typically has a direct cost and perhaps even an opportunity cost. For example, in the case of the ceramic tile manufacturer, one must still pay the fuel cost to keep the kiln burning, the labor and/or machine costs to set dummy tiles on kiln cars that move through the kiln, and so on. These costs are incorporated in the objective functions as $r_{ij}(Y_{ijt} - W_{ijt})$, where $Y_{ijt} - W_{ijt}$ is equal to one when line j is idle during period t and zero in all other cases.

Quite often the shutdown of a production line (*i.e.*, $Y_{ojt} = 1$) involves two types of costs and activities. One type consists of the set of fixed costs and activities required to shut a production line down (*i.e.*, turn off a kiln) and restart a production line. These activities and costs may be thought of as one time actions and the binary variable Z_{ojt} and its associated coefficient q_{oj} reflect these actions. Constraints (3.5) assure that the model assesses these costs only once (*i.e.*, in one period) during a shutdown on a line j. The second type of costs and activities associated with the shutdown of a production line may be thought of as the fixed cost per period (s_{oj}) incurred while the line is shutdown. For example, a down production line may still require periodic maintenance to assure its successful restart in some future period. Also, the shutdown of a production line may result in temporary furloughs and workmen's compensation payments or some other directly related cost or opportunity cost. Typically, this type of cost is incurred each period in a relatively fixed amount. The binary variable Y_{oit} and its associated coefficient s_{oj} reflect this periodic cost.

Constraints (3.7) establish a limit on the maximum number of consecutive periods that a model solution can allow a production line j to sit idle or nonproductive. Constraints (3.7) are optional management policy constraints. These constraints allow the scheduler to specify that a production line cannot indefinitely remain open, but nonproductive, even if cost relationships indicate that this represents a lower cost alternative to shutting a line down. These optional constraints allow the possibility that there may be other factors that are difficult to quantify which dictate that a production line cannot remain open, but nonproductive beyond some period of time. Constraints (3.8) and (3.9) represent additional optional management policy constraints designed to address the same issue as constraints (3.7). Constraints (3.8) limit the total number of periods that any production line j can sit temporarily idle over the entire planning horizon, while constraints (3.9) limit the number of periods that all L lines of a plant collectively can remain idle, but not shutdown, over a planning horizon. Finally, constraints (3.10), in combination with the nonnegativity constraints (3.2) and the integrality constraints (3.11), assure that demand for each product i in period t is satisfied.

The product family plant scheduling model (3.1)-(3.13) illustrates the types of decisions evaluated, and the decision support required at individual plants, at both the tactical and operational planning levels. Utilizing a disaggregation methodology such as that illustrated in section 3.2, a planner can input a network-wide planning model's production plan for an individual plant into model (3.1)-(3.13). This plant level model can then evaluate the feasibility of the tactical production plan taking into consideration such factors as the impact of weekly line changeovers. As just described, this model can also provide input to tactical (annual) decisions such as whether to temporarily shut down or idle a production line that the network-wide model indicates will not need to operate at full capacity.[6] At the operational planning level, model (3.1)-(3.13) can function as a weekly product family scheduling model (see Figure 3.4). In this capacity, it can schedule a plant at the family level for short run planning purposes (e.g., scheduling lines by week for 13 or so weeks in advance). Thus, model (3.1)-(3.13) is representative of the type of versatile decision support models that facilitate effective plant production planning in a hierarchical production planning system. Next, appendix 3.3.2 describes how one can also use model (3.1) – (3.13) to plan periods of production with reduced output rates. Then in section 3.3.3, we will consider an example of how product family production schedules generated by models such as (3.1)-(3.13) must undergo further evaluation at the end item level in order to verify the feasibility of the family level plans.

Appendix 3.3.2 The Equivalence Between Idle Periods And Reduced Production Rates

The purpose of the production scheduling model (3.1)-(3.13) is to assist a manufacturer in deciding whether to plan: (1) the temporary shutdown of production line(s), (2) the idling of production line(s), or (3) a combination of both strategies in response to forecast demand levels that do not fully utilize the plant's available production capacity.

A solution to this model will clearly identify recommended periods to shutdown and/or idle production lines. However, if we again consider the example of a tile manufacturer, a solution to this model also implicitly provides a recommended reduced production rate plan. How a scheduler would translate a solution from this model which recommends some idle production periods into planned reduced production rates requires an explanation.

The scheduler can essentially follow a plan from this model without strictly adhering to a schedule that indicates a line should sit idle, but should not be shut down. For example, suppose that a plan shows that a line should sit idle for 2 weeks and that the line should produce 3 weeks of a product at the regular production rate of 50,000 units per week. One can also interpret this as 5 weeks of production of a product with the 2 weeks of idle time (*i.e.*, firing slug or dummy

---

[6] This illustrative model can also address a third alternative to temporarily shutting down or idling a line, namely, the alternative of running a line at a reduced rate (see appendix 3.3.2).

tiles) interspersed over the 5 weeks. This could translate into 5 weeks of production at a reduced rate of 30,000 units per week rather than 3 weeks at 50,000 units each week. Table 3.4 illustrates how a production scheduler can convert a model solution with idle periods into a planned schedule with reduced rates. In practice, kilns would continue to fire 50,000 units each week (30,000 units of real tile and 20,000 units of slug tile) during the 5-week period. The costs of putting the slug tile through the kiln are still correctly captured by the model. Note that there would be a relatively minor change in the overall inventory carrying costs of the slightly modified production schedule. Specifically, the inventory carrying costs would increase slightly because the modified plan (with a reduced production rate) would produce inventory further in advance of demand than the original plan. However, for a ceramic tile manufacturer, this incremental inventory holding cost is generally minimal. Additionally, the individual period demands will also be satisfied by such an approach. (Obviously one must intersperse the idle period(s) over the periods with positive production *following* the idle periods to assure that demand constraints remain satisfied.) This represents one example of how to translate planned idle periods into production at reduced rates. Naturally, one also has the option to follow exactly the model's solution. (The reader interested in modeling methods designed strictly to determine optimal reduced production rates is referred to Silver, 1990.)

At a more general level, one can also take the total production of all real products (i=1,...,N) scheduled by a model solution for a production line over a planning horizon and divide this by the number of periods in which the model schedules the production line up (*i.e.*, producing or idle). This will yield an actual average production rate per period, which can be compared to the plant's design capacity (based on the scheduled production mix), to determine the de facto reduced production rate if periods of production of slug tiles (*i.e.*, idle periods) have been scheduled.

Table 3.4 The Equivalence Between Idle Periods And Reduced Production

Schedule with Idle Periods			**Equivalent Schedule with Reduced Rates***	
Period	Product	Production Rate	Product	Production Rate
1	1	20,000	1	20,000
2	1	20,000	1	20,000
3	Idle	---	2	30,000
4	Idle	---	2	30,000
5	2	50,000	2	30,000
6	2	50,000	2	30,000
7	2	50,000	2	30,000
•	•	•	•	•
•	•	•	•	•
52				

*There would be a minor change in the overall inventory carrying cost of the translated solution.

3.3.3 Illustrative End Item To Product Family Feedback Loop

Model (3.1)-(3.13) creates a weekly production schedule at the product family level for a plant. In this section, we consider an example of a test at the end item level that a planner can perform to evaluate the feasibility of the weekly product family schedule.

Table 3.5 displays a weekly product family production schedule for a plant. To simplify the illustration, we assume that this plant has just one production line.

Table 3.5 Illustrative Weekly Product Family Production Schedule

Product Family	Planned Total Weeks of Production	Production Week
A	1	1
B	6	2-4, 10-12
C	5	5-9
•	•	•
•	•	•
•	•	•

Note in Table 3.5 that the production plan calls for one week of production of product family A, six weeks of production of product family B, and so on during the planning horizon. We now review an example of a feasibility check of this family schedule at the end item level.

Feasibility Check

Consider the planned production of one week of product family A during period 1. To determine the feasibility of this plan, we must evaluate whether we can schedule the end item production requirements of family A. To illustrate the analysis required, we make the following two additional assumptions:

1. Product family A contains 20 end items, and
2. The plant has a minimum run quantity of a half (1/2) of a day of production for any end item.[7]

Figure 3.7 illustrates two possible outcomes of the feasibility check of the plan to produce one week of family A. In case 1, the production demands for family A are concentrated in end items 1, 2 and 3 only. There exist no production requirements for end items 4 through 20. In this case, the plant can feasibly produce the net requirements for family A in week 1 with the following schedule:

- 2 days of production of end item 1
- 2 days of production of end item 2
- 1 day of production of end item 3

[7] For this illustration, we assume that changeover issues and/or other production economics factors necessitate this minimum.

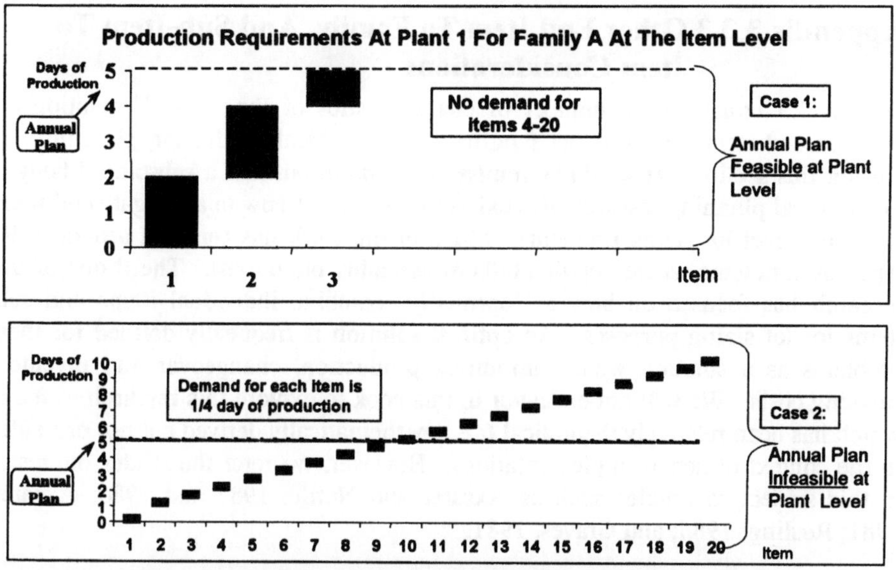

Figure 3.7 Feasibility Checks At End Item Level: Two Cases

In case 2, however, Figure 3.7 illustrates an entirely different situation. Here, the one week of net requirements of family A consists of 1/4 of a day's quantity of production of each of the twenty end items in family A.[8] Thus, the plant must produce all 20 end items in family A. Given the plant's restriction that the minimum run quantity of any end item is 1/2 of a day of production, this means that the plant must actually produce 20 items for 1/2 of a day each. This translates into two weeks (20 x 1/2) required production of product family A. Thus, the original plan created at the product family level[9] is not feasible at the end item level in case 2. In such a situation, this infeasibility which surfaced at the end item level must be communicated back to the product family planning level so that the original plan can be modified to address this infeasibility. Note that this represents just one simple example of the types of infeasibilities that may only become visible at a relatively low level in the production planning and scheduling hierarchy. Naturally, a key to effective planning and scheduling is to recognize the different types of situation dependent infeasibilities that can exist and to address all appropriate infeasibilities. Further, one also needs to determine what, if any, "relatively" minor potential infeasibilities can be ignored in the planning process. While checks and feedback loops are essential to the hierarchical planning process, quite often they do also add time and/or resource requirements to the planning process. Thus, each potential check and feedback loop should be evaluated to determine the importance of its contribution to an effective planning process.

---

[8] ¼ x 20 = 5 days (or one week) of family A net requirements.
[9] By a model such as model (3.1) – (3.13).

Appendix 3.3.3 Other End Item To Family, And Sub-Item To Item Considerations

Section 3.3 considered a number of characteristics of the "linked" relationship between end items and product families in hierarchical production planning. In closing this section, we would be remiss not to point out that a substantial body of hierarchical planning research also exists in the area of how to aggregate end items for multi-level lot sizing problems. Much of the work has centered around MRP type environments where detailed bills of assembly often exist. The thrust of this research has focused on how to "optimally" combine individual items and sub-items for lot sizing purposes. An optimal solution is frequently defined for these problems as a solution which minimizes production, changeover and inventory carrying costs. We will not endeavor in this book to explore this research, some of which has been relatively theoretical (*i.e.*, mathematically derived but not presented in the context of actual implementation). However, we refer the reader interested in this subject to articles such as Axsater and Nuttle, 1987 and 1986; Axsater, 1981; Rosling, 1985; and Graves, 1981.

3.4 Integrated Multi-Echelon Plant Scheduling

The discussion in Chapter 3 through section 3.3 has primarily focused upon hierarchical production planning and scheduling at the individual plant level. In this section, we present a different type of hierarchical plant scheduling, namely, synchronized production scheduling on a multi-echelon manufacturing network where intermediate plants produce work-in-process (WIP) goods and finishing plants transform WIP goods into end products. This represents a different form of operational hierarchical scheduling because the "hierarchy" here lies in the "multi-echelon" dimension. In contrast, the hierarchies that we have previously reviewed at the individual plant level consist of hierarchies of product aggregation, planning horizon and in general, level of detail. In the multi-echelon illustrative scheduling model presented in this section, the planning horizon, product aggregation and level of detail are the same across multiple plants.

Background

The simultaneous (or synchronized) production scheduling of two or more plants represents a challenging hierarchical production scheduling problem. One can add to the complexity and to the value of the solution of such problems by also integrating the scheduling of transportation between an intermediate plant and a finishing plant with the synchronizing of production schedules between two plants. In particular, when multiple transport modes with differing costs and transit times represent viable alternatives, developing synchronized production and transportation schedules which evaluate and optimize all cost and capacity trade-offs is both very difficult and very productive from a practical point of view. The plant scheduling model that follows shortly is designed to illustrate the types of decisions that should ideally be considered in a multi-echelon manufacturing environment.

A multi-echelon production and transportation process includes numerous costs, decision variables and capacities which a firm should evaluate in formulating its production and distribution schedules. These factors can include (but are not limited to) production, changeover and inventory carrying costs at both intermediate and finishing plants, and transportation and intransit inventory carrying costs. Capacity considerations can include production possibilities and constraints at each plant, and in some cases, limits on available transportation capacity. The multitude of costs and decisions in a multi-echelon scheduling process yields a complex problem with a wide variety of implicit interwoven cost and service trade-offs. These trade-offs include balancing long production runs which minimize changeover requirements versus inventory carrying costs, balancing the lower transportation costs associated with slower transport modes versus the resulting increased inventory investment requirements, and so on. In this section, we present a multi-echelon plant scheduling model which simultaneously develops integrated production and transportation schedules for a finishing plant, its supplying intermediate plant and the transport links between the two plants. This optimization model considers all of the costs and production capacities just noted while developing a cost minimizing integrated schedule. It serves to illustrate one approach to hierarchical, multi-echelon scheduling as well as to highlight the issues involved in this type of scheduling, regardless of the methodology employed.

This production and transportation scheduling model represents a short run decision support tool. A firm typically would employ this type of model to assist in scheduling production and transportation operations over at most the next one to two quarters. The scheduling time buckets used will vary by industry and manufacturing process, however, weekly time buckets represent the most common time periods employed. In the scheme of hierarchical production planning, therefore, this production and transportation scheduling model falls into the lowest level of the planning hierarchy, the operational level.

Problem Statement

The following multi-echelon production and transportation mode scheduling model minimizes the total costs to manufacture work-in-process (WIP) goods at an intermediate plant, ship WIP materials from an intermediate plant to a finishing plant, and then manufacture finished goods inventory (FGI) at the finishing plant. Projected demand for finished goods inventory of product families (or end items) on a period-by-period basis represents the "driving" constraint of this model. A solution to this model produces a line-by-line and product family by product family production schedule for each period in a planning horizon for both the intermediate plant(s) and its finished plant(s). Additionally, a model solution generates a transportation mode schedule, by product family, for shipments between the intermediate plant and the finishing plant over the entire planning horizon. The specific costs evaluated by this model include: variable manufacturing costs and changeover costs at each plant, inventory carrying costs at each plant (WIP at the intermediate plant, and WIP and FGI at the finishing plant), in-transit WIP inventory carrying costs, and transportation costs. Finally the following formulation does not allow backlogging of demand.

Production and Transportation Mode Scheduling Model

Note that in the following formulation plant 1 produces intermediate or WIP products, and plant 2 produces finished products. Now consider the following notation in Table 3.6.

Table 3.6 Notation For The Production And Transportation Mode Scheduling Problem

N	is the number of intermediate or (WIP) products at plant 1
N'	is the number of intermediate or (WIP) products at plant 2
N''	is the number of finished products at plant 2
L	is the number of lines at plant 1
L''	is the number of lines at plant 2
T	is the number of periods (in weeks)
i	is the index on the intermediate products ($i = 0,1,\ldots,N$) at plant 1 and the intermediate products at plant 2 ($i = 1,2,\ldots N'$), and the finished products at plant 2 ($i = 0,1,\ldots,N''$). Product zero is not a real product. If a production line j is assigned to product zero during period t, this means that j is idle during period t.
j	is the index on the production lines or processors ($j = 1,\ldots,L$) at plant 1 and ($j = 1,\ldots, L''$) at plant 2
t	is the index on the production period ($t = 1,\ldots, T$)
L_i	is the set of lines that can produce product i (at plant 1)
L_i''	is the set of lines that can produce product i (at plant 2)
M_j	is the set of WIP products that can be produced on line j at plant 1
M_j''	is the set of FGI products that can be produced on line j at plant 2
c_{ij}	is the unit variable manufacturing cost to produce product i on line j at plant 1
c_{ij}''	is the unit variable manufacturing cost to produce product i on line j at plant 2
h_i	is the unit carrying cost per period to hold WIP product i at plant 1
h_i'	is the unit carrying cost per period to hold WIP product i at plant 2
h_i''	is the unit carrying cost per period to hold finished product i at plant 2
q_{ij}	is the fixed cost to changeover to product i on line j, for $i = 1,\ldots,N$ at plant 1
q_{ij}''	is the fixed cost to changeover to product i on line j, for $i = 1,\ldots,N''$ at plant 2
p_{ij}	is the fixed production quantity per period of product i on line j at plant 1
p_{ij}''	is the fixed production quantity per period of product i on line j at plant 2
d_{it}	is the demand for finished product i in period t
II_{io}	is the starting inventory of WIP product i at plant 1
II_{io}'	is the starting inventory of WIP product i at plant 2
IF_{io}	is the starting inventory of finished product i at plant 2
$IIMAX_{it}$	is the maximum inventory level allowed (or upper inventory limit) of WIP product i at the end of period t at plant 1

Table 3.6 Notation For The Production And Transportation Mode Scheduling Problem (continued)

$II'MAX_{it}$	is the maximum inventory level allowed of WIP product i at the end of period t at plant 2
$IFMAX$	is the maximum inventory allowed of finished product i at the end of period t at plant 2
$IIMIN_{it}$	is the minimum inventory level allowed (or lower inventory target) for WIP product i at the end of period t at plant 1
$II'MIN_{it}$	is the minimum inventory level allowed (or lower inventory target) for WIP product i at the end of period t at plant 2
$IFMIN$	is the minimum inventory level allowed (or lower inventory target) for finished product i at the end of period t at plant 2
II_{it}	is the end of period WIP inventory of product i in period t at plant 1
II'_{it}	is the end of period WIP inventory of product i in period t at plant 2
IF_{it}	is the end of period finished goods inventory of product i in period t at plant 2
W_{ijt}	$= \begin{cases} 1, & \text{if WIP product } i \text{ is produced on line } j \text{ in period } t \text{ at plant 1} \\ 0, & \text{otherwise} \end{cases}$
W''_{ijt}	$= \begin{cases} 1, & \text{if finished product } i \text{ is produced on line } j \text{ in period } t \text{ at plant 2} \\ 0, & \text{otherwise} \end{cases}$
Z_{ijt}	$= \begin{cases} 1, & \text{if production line } j \text{ changes to product } i \text{ in period } t \text{ at plant 1} \\ 0, & \text{otherwise} \end{cases}$
Z''_{ijt}	$= \begin{cases} 1, & \text{if production line } j \text{ changes to product } i \text{ in period } t \text{ at plant 2} \\ 0, & \text{otherwise} \end{cases}$
U_{it}	is the quantity of WIP product i shipped from plant 1 to plant 2 by the fastest mode (e.g., air) in shipments originating during period t
a_{it}	is the unit cost to transport WIP product i by the fastest mode from plant 1 to plant 2 in shipments originating during period t
V_{it}	is the quantity of WIP product i shipped by the second fastest mode from plant 1 to plant 2 in shipments originating during period t
b_{it}	is the unit cost to transport WIP product i by the second fastest mode from plant 1 to plant 2 in shipments originating during period t
X_{it}	is the quantity of WIP product i shipped by the slowest mode (e.g., surface) from plant 1 to plant 2 in shipments originating during period t
s_{it}	is the unit cost to transport WIP product i by the slowest mode from plant 1 to plant 2 in shipments originating during period t
f_i	is the unit intransit inventory carrying cost for WIP product i when shipped by the fastest mode from plant 1 to plant 2
g_i	is the unit intransit inventory carrying cost for WIP product i when shipped by the second fastest mode from plant 1 to plant 2
e_i	is the unit intransit inventory carrying cost for WIP product i when shipped by the slowest mode from plant 1 to plant 2

The production/transportation model is as follows:

Minimize:
$$\sum_{t=1}^{T}\sum_{i=1}^{N}\sum_{j\in L}\left(c_{ij}p_{ij}W_{ij}+q_{ij}Z_{ijt}\right)+\sum_{t=1}^{T}\sum_{i=1}^{N}h_iII_{it}$$
$$+\sum_{t=1}^{T}\sum_{i=1}^{N''}\sum_{j\in L''}\left(c''_{ij}p''_{ij}W''_{ijt}+q''_{ij}Z''_{ijt}\right)+\sum_{t=1}^{T}\sum_{i=1}^{N'}h'_iII'_{it}$$
$$+\sum_{t=1}^{T}\sum_{i=1}^{N''}h''_iIF_{it}+\sum_{t=1}^{T}\sum_{i=1}^{N}\left(a_{it}U_{it}+f_iU_{it}\right)$$
$$+\sum_{t=1}^{T}\sum_{i=1}^{N}\left(b_{it}V_{it}+g_iV_{it}\right)+\sum_{t=1}^{T}\sum_{i=1}^{N}\left(s_{it}X_{it}+e_iX_{it}\right) \qquad (3.14)$$

Subject to:

$II_{it} \geq IIMIN_{it}$	$i = 1,\ldots, N$	$t = 1,\ldots, T$	(3.15)
$IT_{it} \leq IIMAX_{it}$	$i = 1,\ldots, N$	$t = 1,\ldots, T$	(3.16)
$II'_{it} \geq II'MIN_{it}$	$i = 1,\ldots, N'$	$t = 1,\ldots, T$	(3.17)
$II'_{it} \leq II'MAX_{it}$	$i = 1,\ldots, N'$	$t = 1,\ldots, T$	(3.18)
$IF_{it} \geq IFMIN_{it}$	$i = 1,\ldots, N''$	$t = 1,\ldots, T$	(3.19)
$IF_{it} \leq IFMAX_{it}$	$i = 1,\ldots, N''$	$t = 1,\ldots, T$	(3.20)

{Demand for FGI}

$$IF_{it} = IF_{it-1} + \sum_{j\in L''}\sum_{k=1}^{t} p_{ij}W''_{ijk} - \sum_{k=1}^{t} d_{ik}; \qquad i = 1,\ldots,N'';\ t = 1,\ldots,T \qquad (3.21)$$

{WIP at Plant 1}

$$II_{it} = II_{it-1} + \sum_{j\in L}\sum_{k=1}^{t} p_{ij}W_{ijk} - U_{it} - V_{it} - X_{it}; \qquad i = 1,\ldots,N;\ t = 1,\ldots,T \qquad (3.22)$$

{WIP at Plant 2}

$$II'_{it} = II'_{it-1} + U_{it-1} + V_{it-2} + X_{it-3} - \sum_{j\in L''}\sum_{k=1}^{t} p''_{ij}W''_{ijk}; \qquad i = 1,\ldots,N';\ t = 1,\ldots,T \qquad (3.23)$$

{Production of FGI ≤ WIP}

$$\sum_{j\in L''}\sum_{k=1}^{t} p''_{ij}W''_{ijk} \leq II'_{it-1} + U_{it-1} + V_{it-2} + X_{it-3}; \qquad i = 1,\ldots,N';\ t = 1,\ldots,T \qquad (3.24)$$

{Changeover at Plant 1}

$$Z_{ijt} \geq W_{ijt} - W_{ijt-1}; \qquad j\varepsilon L_i;\ i = 0,\ldots,N;\ t = 1,\ldots,T \qquad (3.25)$$

{Changeover at Plant 2}

$$Z''_{ijt} \geq W''_{ijt} - W''_{ijt-1}; \qquad j\varepsilon L''_i;\ i = 0,\ldots,N'';\ t = 1,\ldots,T \qquad (3.26)$$

$$\sum_{i\in M_j} W_{ijt} = 1; \qquad j = 1,\ldots,L; \qquad t = 1,\ldots,T \qquad (3.27)$$

$$\sum_{i\in M_j''} W_{ijt}'' = 1; \qquad j = 1,\dots,L''; \qquad t = 1,\dots,T \tag{3.28}$$

$$0 \le W_{ijt} \le 1 \text{ integer}; \qquad j\varepsilon L_i; \qquad i = 0,\dots,N; \qquad t = 1,\dots,T \tag{3.29}$$

$$0 \le W_{ijt}'' \le 1 \text{ integer}; \qquad j\varepsilon L_i''; \qquad i = 0,\dots,N; \qquad t = 1,\dots,T \tag{3.30}$$

$$Z_{ijt} \ge 0; \qquad j\varepsilon L_i; \qquad i = 0,\dots,N; \qquad t = 1,\dots,T \tag{3.31}$$

$$Z_{ijt}'' \ge 0; \qquad j\varepsilon L_i''; \qquad i = 0,\dots,N''; \qquad t = 1,\dots,T \tag{3.32}$$

$$IIMIN_{it} \ge 0; \qquad i = 1,\dots,N; \qquad t = 1,\dots,T \tag{3.33}$$

$$II'MIN_{it} \ge 0; \qquad i = l,\dots N'; \qquad t = 1,\dots,T \tag{3.34}$$

$$IFMIN \ge 0; \qquad i = l,\dots,N''; \qquad t = 1,\dots,T \tag{3.35}$$

$$U_{it}, V_{it}, X_{it} \ge 0; \qquad i = 1,\dots,N'; \qquad t = 1,\dots,T \tag{3.36}$$

Constraints (3.15) and (3.16) assure that the end of period WIP inventories at plant 1 (the intermediate echelon plant) are greater than or equal to the minimum inventory required for each product family *i*, but do not exceed the maximum period targets for each product family. Constraints (3.17) and (3.18) perform the same function for WIP inventories at plant 2 (the second echelon finishing plant) – *i.e.*, they establish an allowable end of period inventory target range for each product family for each period of the planning horizon. Constraints (3.19) and (3.20) collectively set the end of period inventory target ranges for each family of finished goods at plant 2. The non-negativity constraints (3.33), (3.34) and (3.35) on *IIMIN*, *II'MIN*, and *IFMIN* respectively, assure that backlogging of demand cannot occur. Finally, establishing maximum inventory levels for each product family through constraints (3.16), (3.18) and (3.20) assures that model solutions do not produce excessively high inventories to avoid the potentially high costs of idling a production line. Whether this would occur in the absence of these constraints depends upon the comparative costs associated with idling a line, and producing and carrying excess inventory. (The costs of idling a line for a period can be specified through the value of the cost coefficient associated with a period of production of product 0 on a line at either plant 1 or plant 2.)

Constraints (3.21) in combination with non-negativity constraints (3.35) and integrality constraints (3.28) ensure that model solutions satisfy demand for each FGI family *i* in each period *t* at plant 2. Constraints (3.22), (3.33) and (3.27) in a similar fashion assure that all demands for each WIP family during each period at plant 1 are satisfied. Finally, constraints (3.23), (3.34) and (3.28) collectively perform this same function with respect to the demand for each WIP family during each period at plant 2. Constraints (3.22) and (3.23) in effect also describe the flow of WIP families between plant 1 and plant 2. Constraints (3.22) show that the end of period inventory of a family *i* at intermediate plant 1 represents the sum of the family's beginning of period inventory level, plus the production of the family during that period, less the "produced" WIP inventory of the family shipped to plant 2 during the period. Notice that equations (3.22) have three variables to represent outbound shipments: U_{it} for the "fastest" available transport mode between the plants, V_{it} for the "second fastest" mode and X_{it} for the "slowest" mode. The respective speeds of the alternative transport modes available for interplant shipments is indicated in constraints (3.23) by the time lags shown on variables U_{it}, V_{it} and X_{it}. Constraints (3.23) indicate that using the fastest mode of transport, WIP becomes available at plant 2 the next period after it is shipped from plant 1, while when the second fastest and slowest modes of transport are employed, shipments of WIP inventory do not become available at plant 2 until two and three periods in the future respectively. In this formulation, we have implicitly assumed that whatever quantity of WIP plant 1 ships to plant 2 by a particular mode during a period t arrives at plant 2 in full, at or by the beginning of a future period. Thus, the formulation cannot model with precise accuracy a situation where the relationship between transportation leadtimes, shipping schedules and production periods is such that WIP arrives at plant 2 in the middle of a production period t and could theoretically be used as input for production of FGI for some fraction of that period t. It is relatively easy to construct an alternative formulation that can address this situation and the reader interested in this is referred to De Matta and Miller (2000). Note also that one can easily adjust both the time lags for flows of WIP between plants 1 and 2 by alternative modes, as well as the number of modes (variables) available. In this formulation we presume three transport mode choices based upon some real world examples. For example, the fastest mode alternative consist of air shipments between two plants, the second fastest alternative may represent intermodal air-surface moves, and the slowest alternative may represent all surface moves (e.g. ocean-local truck drayage).

Constraints (3.24) stipulate that the finishing plant cannot produce more FGI of a family i during period t than it has WIP of that family to produce. Note that for purposes of simplicity, constraints (3.24) indicate that to produce one unit of FGI of a family i requires one input unit of WIP of family i. However, one can easily adjust constraints (3.24) and (3.23) to reflect manufacturing environments where the relationship between WIP and FGI inventory is not one to one. For example, assume that we created a parameter m for each product family i which indicates how many units of FGI could be produced from one unit of WIP inventory of i (e.g., 0.5). By replacing $p''_{ij}W''_{ijk}$ with $m''_{ij}p''_{ij}W''_{ijk}$ in constraints (3.24), and $p''_{ij}W''_{ijk}$ with $(1/m''_{ij}p''_{ij}W''_{ijk})$ in constraints (3.23), we could describe alternative manufacturing relationships between WIP and FGI inventory. Similarly, one could also adjust our proposed model to depict manufacturing processes where multiple WIP product families are required as inputs to produce one FGI family. Finally, note that constraints (3.24) again reflect the different shipment speeds offered by alternative transport modes between plants 1 and 2.

Constraints (3.25) relate the changeover variable (Z_{ijt}) to the WIP production variables in the current and previous periods (W_{ijt} and W_{ijt-1} respectively) at plant 1. A changeover to product i on line j occurs during period t, if and only if $W_{ijt} = 1$ and $W_{ij\,t-1} = 0$. It is easy to demonstrate that constraints (3.25) force Z_{ijt} to be one only if $W_{ijt} = 1$ and $W_{ijt-1} = 0$; in all other cases Z_{ijt} must be zero. Therefore, Z_{ijt} need not explicitly be specified as a binary variable. Constraints

(3.26) stipulate this same relationship between the changeover variable and the FGI production variables of the current and previous periods on line j at the finishing plant (2). One may note that this formulation includes changeover variables, but does not include explicit "line set-up" variables. This differs from models such as (3.1) – (3.13) which have separate variables to represent line set-up costs and line changeover costs. As previously noted in section 3.2, the scheduling model (3.1) – (3.13) focuses on evaluating whether a production line should be shut down or temporarily idled when insufficient demand exists to fully consume the capacity of a plant's lines. In order to examine the cost trade-offs associated with idling versus shutting a line down, it necessary to explicitly define both line set-up and line changeover costs. However, this trade-off is not central to the production and transportation mode scheduling model proposed herein. Therefore, for purposes of simplicity, we simply use the "scheduling of product 0" on a line to represent the idling of a line for a period, and we associate a single period cost coefficient to represent the cost of idling a line. (*i.e.*, The cost coefficient q_{oj} represents the total cost to idle a line j for one period.) The more complex analysis of an idling decision available in (3.1) – (3.13) can be incorporated into model (3.14) – (3.36) if appropriate.

Constraints (3.27) and (3.29) assure that each production line at plant 1 produces exactly one product during each period. Recall that a line j producing product 0 during period t ($W_{0jt} = 1$) is idle. Constraints (3.28) and (3.30) similarly constrain each line j at plant 2 to produce exactly one product during period t. Finally, equations (3.31) through (3.36) represent standard non-negativity constraints.

Real World Scheduling Perspective

In practice, most firms have a standard replenishment plan for how their intermediate plants should ship WIP inventory to their finishing plants. Typically, based on cost trade-offs (e.g., freight costs and inventory carrying costs) and other considerations, a firm determines the normal mode for transporting WIP inventory between two plants. Similarly, firms will usually establish normal shipping frequencies between two plants based upon cost trade-offs and/or operational considerations (e.g., sailing schedules in the case of ocean transport). Thus, returning to the hierarchical planning paradigm illustrated in Figure 3.2, during their tactical (annual) planning processes, most firms will have set general guidelines for multi-echelon replenishment between their respective intermediate and finishing plants. Model (3.14) – (3.36) then comes into play at the short run operational level. Employing an integrated approach for simultaneously developing production schedules across two manufacturing echelons and developing the transportation schedule to move WIP inventory between echelons helps determine:

(1) short run production schedules and quantities at each plant(s),
(2) if it necessary to use a different mode than normal for any shipments between plants (*i.e.*, the short run transportation schedule), and

(3) the quantities of each WIP product i to be shipped, by mode, from plant 1 to plant 2 during each period.

In practice, model (3.14) - (3.36) represents a difficult model to solve, although heuristics have been developed to solve such formulations (e.g., De Matta and Miller, 2000). However, this review of a synchronized multi-echelon plant scheduling model has illustrated yet another type of hierarchical planning that must be considered in real world planning. The decisions embodied in (3.14) – (3.36) are very representative of the multi-echelon production planning and scheduling considerations that practitioners must address every day.

3.5 Material Requirements Planning

A review of hierarchical production planning systems would be incomplete without at least some mention of materials requirements planning. Material requirements planning (MRP) represents yet another hierarchical planning technique used to plan and schedule manufacturing operations at a plant level. MRP systems have a strong operational focus and the origins of such systems trace back to such early practitioners as Joseph Orlicky (1975) and Oliver Wight (1981). Briefly, MRP systems are designed to calculate the schedule and the requirements for the production of components and subassemblies, and the purchase of materials such that the finished goods requirements generated by a master production schedule (MPS) can be met. Further a MRP system performs at least three major functions including: (1) determining WIP inventory requirements and controlling WIP inventory levels, (2) prioritizing the production and/or purchase of components, and (3) determining production requirements at a detailed level below the finished goods end item level. Over time MRP has evolved from a plant level hierarchical planning system focused on planning material orders to in many cases a much broader planning system referred to as manufacturing resources planning or MRP II (see e.g., Melnyk et. al., 1983). Today, the domain of MRP II systems can extend to planning all the resources of a manufacturing organization in a closed loop process across a large scale network.

In this section, we will review at a summary level material requirements planning (MRP). The intent here is not to examine MRP in detail, but rather to understand how MRP represents an extension or complement to the hierarchical planning systems reviewed to this point. The reader interested in a detailed review of MRP is referred to Orlicky et. al. (1975), Wight (1981), Plossol (1980), and Melnyk et. al. (1983), among others.

3.5.1 The Role Of MRP In A Production Planning Hierarchy

To clarify the role of MRP in a hierarchical production planning system, we return to the illustrative schematic for hierarchical tactical and operational planning and scheduling originally presented in Chapter 1. Figure 3.8 displays this same hierarchical representation with a very simple MRP system appended to it.

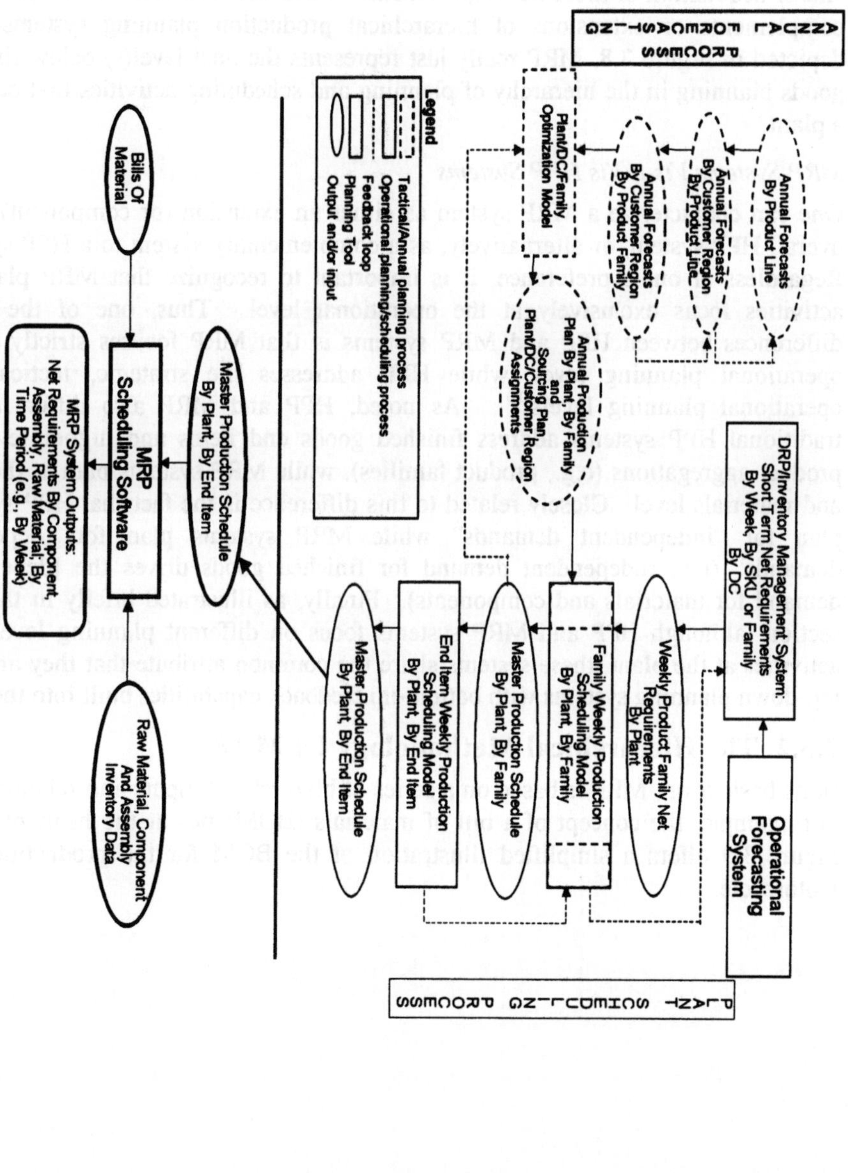

FIGURE 3.8 The Role Of MRP In A Hierarchical Planning System (Integrated Production And Distribution Planning System)

Note that the finished goods end item master production schedule of a plant drives the MRP system. As noted, the MPS defines the net requirements schedule for finished goods at a plant by defined time bucket. (Typically these time buckets are established in weekly increments.) These end item net requirements then drive the net requirements and timing of requirements for the raw materials, components and sub-assemblies used to produce finished goods. The primary role of an MRP system is to calculate and provide planning guidance for the production and/or acquisition of these net requirements. Thus, one can view MRP systems as complements or extensions of hierarchical production planning systems. As depicted in Figure 3.8, MRP really just represents the next level(s) below finished goods planning in the hierarchy of planning and scheduling activities that occur at a plant.

MRP Systems Vis-a-Vis HPP Systems

One can characterize a MRP system as either an extension (or component) of an overall HPP system, or alternatively, as a complementary system to a HPP system. Regardless of one's preference, it is important to recognize that MRP planning activities focus exclusively at the operational level. Thus, one of the many differences between HPP and MRP systems is that MRP focuses strictly at the operational planning level, while HPP addresses the strategic, tactical and operational planning levels.[10] As noted, HPP and MRP also differ in that traditional HPP systems address finished goods end items and higher levels of product aggregations (e.g., product families), while MRP systems plan at the WIP and materials level. Closely related to this difference is the fact that HPP systems plan for "independent demands" while MRP systems plan for "dependent demands" (*i.e.*, independent demand for finished goods drives the [dependent] demand for materials and components). Finally, as illustrated briefly in the next section, although HPP and MRP systems focus on different planning levels and activities at the plant, these systems share the common attribute that they are both top-down planning systems with bottom-up feedback capabilities built into them.

3.5.2 The Hierarchical Methodology Of MRP

In its basic form, MRP is based on a series of hierarchical inputs and relationships. For example, the concept of a bill of materials (BOM) lies at the heart of MRP. Figure 3.9 offers a simplified illustration of the BOM for the production of a motorcycle.

[10] Note that full scale MRPII systems can provide decision support at the tactical level.

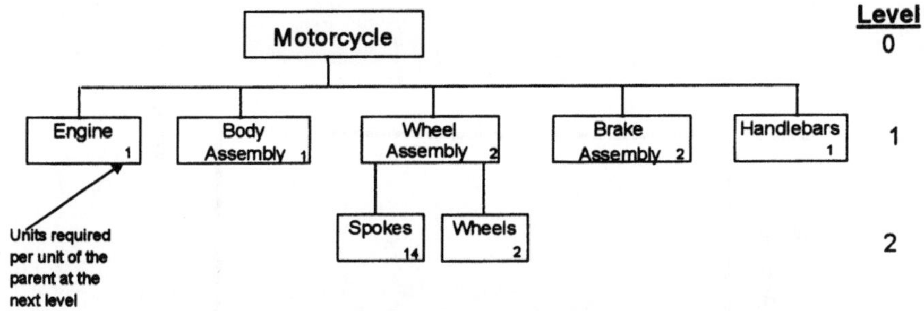

Figure 3.9 Illustrative Bill Of Material For A Motorcycle

As this figure illustrates, the BOM essentially represents a hierarchical description of how raw materials, components and/or sub-assemblies must be put together in order to produce a finished product. The BOM provides critical information on both the sequences and quantities required to complete one unit of a finished end item. Another important element of a MRP system consists of the lead time offset data or the time-phased assembly schedule. Figure 3.10 displays a simplified set of lead time offsets for the production of a motorcycle. The lead time offsets or time phased assembly schedules in a MRP system add the element of timing to the sequence and component quantity data provided by the BOM. Again, the lead time offset table has a hierarchical dimension to it, reflecting the hierarchical process of manufacturing a finished good end item. The intent of these illustrations is not to provide expansive detail about MRP, but rather to note the use of a hierarchical planning concepts in MRP systems. Thus, we refer the reader to previous MRP citations for more detailed explanations of this technique.

Finally, as noted previously, practitioners and developers have greatly expanded the basic capabilities of MRP systems over the last two decades, and MRP II quickly followed the development of MRP. Although a detailed discussion of MRP II is outside of the objectives of this section, it is instructive to briefly consider some extensions to basic MRP.

The technique of MRP does not address capacity issues directly. MRP identifies the net materials requirements necessary to produce a finished goods schedule. However, it does not evaluate the feasibility of producing the calculated net requirements in terms of the available capacity. To address capacity issues at the materials, sub-assemblies and components level, MRP developers and practitioners created a technique called capacity requirements planning (CRP). CRP performs a detailed analysis of the labor and machine requirements necessary to produce a net requirements schedule generated by a MRP calculation. Briefly, CRP develops these analyses at a detailed work center level. A key output of a CRP system is a report or series of reports (often termed a "load" report) that calculates detailed capacity requirements by work center and materials. These reports identify any situations where insufficient capacity exists to meet the net requirements of the MRP schedule.

The addition of techniques such as CRP to the MRP process led to what is known as closed loop MRP. Figure 3.11 offers a very rudimentary illustration of a closed loop MRP system.

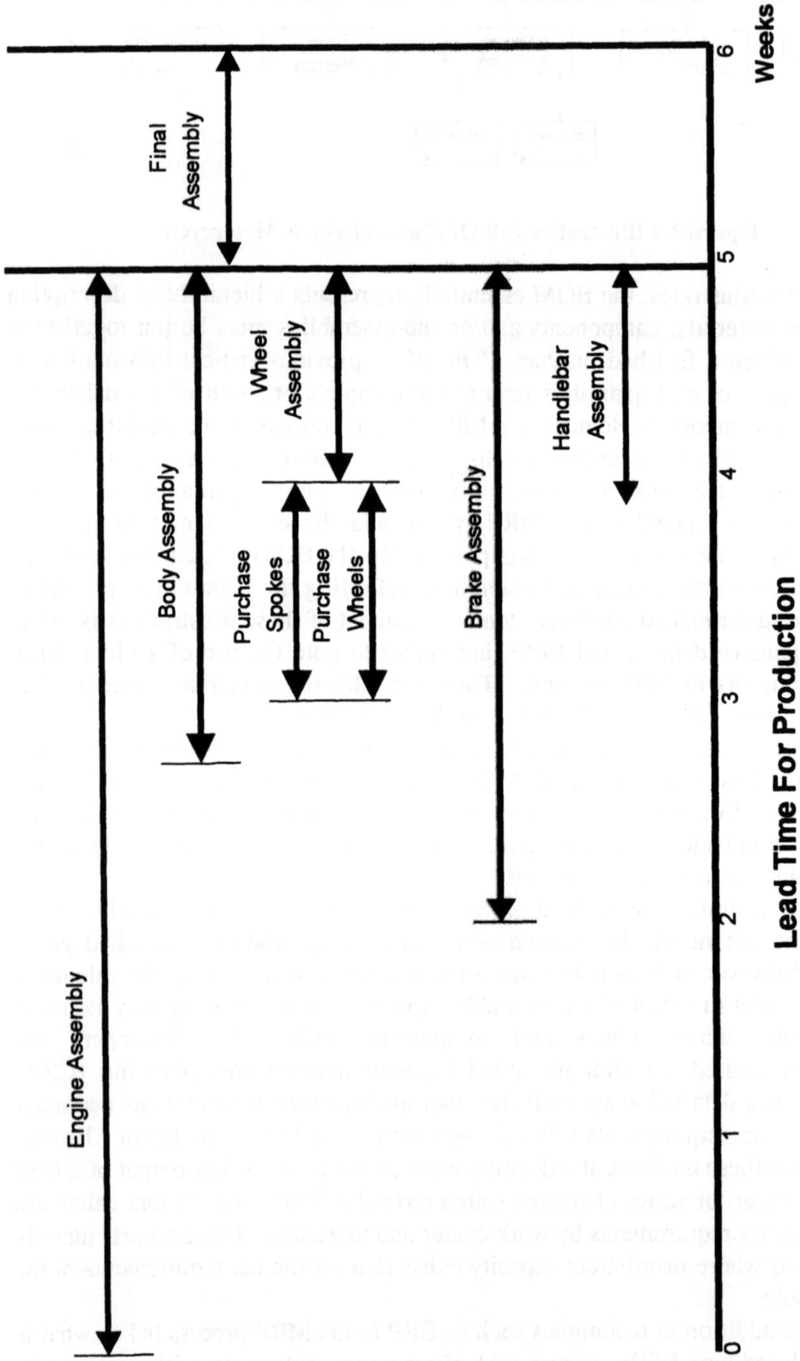

Figure 3.10 Illustrative MRP Lead Time Offsets

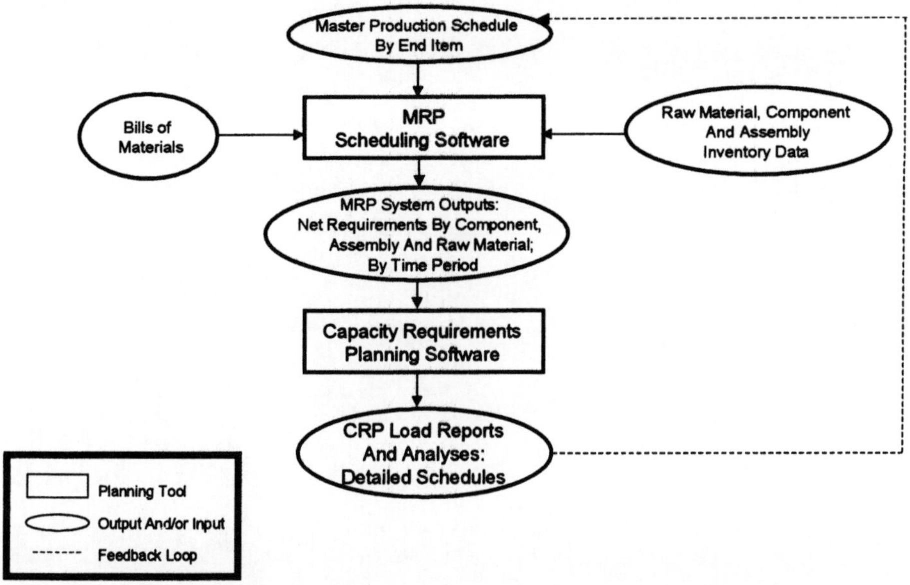

Figure 3.11 Illustrative MRP And CRP Closed Loop System

Note that similar to the HPP systems previously reviewed, a closed loop MRP system contains a feedback loop to communicate potential infeasibilities (e.g., capacity constraint violations). In the illustration in Figure 3.11, if the capacity requirements planning evaluation of the MRP net requirements schedule indicates an infeasibility, a feedback loop flows back to the original master production schedule. This feedback loop would communicate the infeasibility to a planner, and then typically, a planner would determine the adjustments required to yield a feasible MPS and MRP net requirements plan. Thus, the feedback loop from the CRP module serves to assure that the final MRP production schedule developed does in fact represent a feasible plan.

3.6 Summary

In this chapter, we have reviewed a broad spectrum of hierarchical production planning techniques designed to provide decision support at the plant production planning and scheduling level. These methodologies ranged from an optimization based product family scheduling model approach to evaluate tactical network-wide plant production plans at the individual plant level, to techniques such as MRP to schedule production at the component and assembly level. The planning and scheduling areas discussed in this chapter have covered only a portion of the hierarchical production planning techniques used today at the individual plant level. The cross section reviewed in this chapter, however, does serve to illustrate the importance of applying hierarchical techniques in strategic, tactical and operational production planning and scheduling at the plant level.

Chapter 4

Distribution And Transportation Planning And Scheduling

Chapter 4 explores hierarchical planning and scheduling activities and techniques for two integral components of the supply chain; namely, distribution and transportation. As in Chapter 3, we again assume that fundamental decisions regarding the planning and utilization of the network have been developed in an integrated manufacturing and distribution approach (using methods such as those considered in Chapters 1 and 2), and the firm's attention in this chapter turns more towards stand-alone distribution and transportation questions. Further, in making this assumption, it is important to note that in real world practice, many firms do not employ a true "integrated" manufacturing and distribution approach in the planning of a network's infrastructure or in the execution of a network's operations.[1] Instead, manufacturing and distribution often develop and execute their strategies as relatively stand-alone functional entities. For many planning problems and decisions, this represents an appropriate decomposition approach that simplifies a complex set of issues or decisions into a more manageable subset. In other cases, this approach can have significant, undesirable ramifications for a firm, and the myriad reasons why such practices occur could by themselves form the basis for a book. For the purposes of this chapter, however, we simply want to recognize that manufacturing and distribution network planning decisions do not always take place in an integrated approach, and that firms often make strategic distribution infrastructure decisions in isolation. Further, in this chapter, we will provide several examples to illustrate the types of strategic distribution and transportation decisions that firms typically can effectively make in a relatively stand-alone environment.

The first section of Chapter 4 reviews a cross-section of distribution and transportation strategic and tactical infrastructure planning problems found in industry. These problems range from designing and monitoring a large-scale multi-echelon distribution and transportation network, to a strategic rail network design problem, to the development of a direct store delivery (DSD) local depot and delivery fleet operation designed to serve an individual market. The next section (4.2) of this chapter offers an overview of short run transportation planning and scheduling approaches. This will naturally include a review of the classic vehicle routing problem which practitioners have often utilized a hierarchical approach to solve and implement. In section 4.3, we will examine additional techniques for linking short run transportation scheduling and operational activities to tactical/annual transportation planning activities. Then section 4.4 will consider

[1] Further, many firms that supply goods to customers do not themselves have manufacturing operations (e.g., distributors and wholesalers).

another important facet of distribution and transportation planning, namely, transport mode selection. In particular, this section will present a hierarchical planning technique to provide mode selection decision support. This presentation will illustrate the need for simple product aggregation and hierarchy schemes to facilitate effective tactical mode choice selection. Finally, in section 4.5, we discuss approaches for a firm to bundle (*i.e.*, aggregate) its annual transportation shipping requirements to facilitate effective negotiations with third party transport service providers. Negotiations to procure transportation services from providers represent an important element in the overall activities of transportation managers in private industry. An important factor in crafting an effective negotiating strategy is how well a firm aggregates its service requirements into a hierarchy or set of transport bundles to present to providers for bidding purposes. In recent years, mathematical modeling based decision support systems to facilitate this aggregation process have begun to attract attention, and we will briefly consider one such system in section 4.5.

To provide background perspective on the topics addressed in Chapter 4, we start by considering the hierarchy of distribution and transportation network planning issues and decisions outlined in Figures 4.1 through 4.3. A number of these decisions (e.g., new warehouse locations and sizes - in Figure 4.1) previously appeared in Chapter 1[2] under integrated strategic, tactical and operational manufacturing and distribution planning. This illustrates that a number of firms address certain distribution planning decisions in an integrated manufacturing and distribution planning process, while other firms may treat these decisions in a relatively independent manner.

Figures 4.1 through 4.3 present just a small sample of the numerous strategic through operational decisions required on an ongoing basis to facilitate the effective management of a large scale distribution and transportation network. Nevertheless, this small sample serves to illustrate the hierarchical characteristics of network transportation and distribution planning.

- New warehouse locations, sizes and materials handling technologies
- Warehouse closings
- Mix of owned assets vs. third party resources utilized (*i.e.*, outsourcing decisions)
- Design of transportation network and transport mode strategy
- Selection of transportation providers for long term partnerships
- Planned order cycle time standards and capabilities on customer orders
- Planned customer service levels (*i.e.*, order and line fill rates)

Figure 4.1 Selected Strategic Distribution And Transportation Decisions

[2] In Figures 1.1, 1.2 and 1.3 of Chapter 1

- Warehouse workforce requirements
- Warehouse materials handling and storage/racking capacities
- Planned utilization rates of available materials handling technologies
- Transportation fleet size and workforce (if maintain a private fleet)
- Transportation mode and carrier selection
- Longer-term transportation routing

Figure 4.2 Selected Tactical Distribution And Transportation Decisions

- Short-term (distribution) balancing of inventory deployed across a network
- Customer order processing and scheduling
- Warehouse operations scheduling
- Labor scheduling for warehouse and transportation operations
- Vehicle scheduling and short-term routing
- Carrier selection and mode choice for individual loads

Figure 4.3 Selected Operational Distribution And Transportation Decisions

4.1 Strategic And Tactical Distribution And Transportation Planning

In this section, we review several different distribution, transportation infrastructure and mode choice planning problems. We begin with a brief review of a classic distribution strategic planning problem, namely, the warehouse facility location problem. In reality, the warehouse location problem typically is not categorized as a hierarchical planning problem. If there exists a hierarchical dimension to the warehouse location problem, it is simply that such problems can involve planning the locations of multiple facilities on two or more echelons of a distribution network (*i.e.*, a hierarchy of echelons). However, although not hierarchical in nature, the warehouse location problem represents a good starting point for a discussion of strategic distribution planning.

4.1.1 The Warehouse Location Model

Table 4.1 displays the notation that we will use to formulate a warehouse location model.

Table 4.1 Notation For A Warehouse Location Model

$s_{ij\ell}$	is the quantity of product ℓ shipped from plant i to warehouse j, where $i = 1,...,M$; $j = 1,..., N$; and $\ell = 1,...,P$
$t_{jk\ell}$	is the quantity of product ℓ shipped from warehouse j to demand region k, where $k = 1,...,R$
$d_{k\ell}$	is the demand (in units) of region k for product ℓ
y_j	$= \begin{cases} 1 \text{ if warehouse } j \text{ is opened} \\ 0 \text{ otherwise} \end{cases}$
w_{jk}	$= \begin{cases} 1 \text{ if warehouse } j \text{ supplies region } k \\ 0 \text{ otherwise} \end{cases}$
$a_{ij\ell}$	is the cost per unit to ship product ℓ from plant i to warehouse j
$b_{jk\ell}$	is the cost per unit to ship product ℓ from warehouse j to demand region k
F_j	is the annual fixed cost to operate warehouse j
$v_{j\ell}$	is the variable cost per unit to move product ℓ through warehouse j
I_j	is the one-time investment cost required to build or open a new warehouse j
$C_{i\ell}$	is the annual production capacity of plant i to manufacture product ℓ
U	is the maximum number of warehouses that the firm wants to open or operate
B	is the maximum financial investment in opening new warehouses that the firm wants to make (*i.e.*, a budget constraint)

A Warehouse Location Model

Minimize:

$$\sum_{i=1}^{M}\sum_{j=1}^{N}\sum_{\ell=1}^{P} a_{ij\ell}s_{ij\ell} + \sum_{j=1}^{N}\sum_{k=1}^{R}\sum_{\ell=1}^{P}\left(b_{jk\ell}t_{jk\ell} + w_{jk}v_{j\ell}d_{\ell k}\right)$$

$$+\sum_{j=1}^{N}\left(y_jF_j + y_jI_j\right) \tag{4.1}$$

Subject to:

$$\sum_{j=1}^{N} s_{ij\ell} \le C_{i\ell}, \text{ for each } i = 1,...,M \text{ and } \ell = 1,..., P \tag{4.2}$$

$$\sum_{j=1}^{N} t_{jk\ell} - d_{k\ell} = 0 \text{ for each } k = 1,...,R, \text{ and } \ell = 1,...,P \tag{4.3}$$

$$\sum_{j=1}^{N} w_{jk} = 1 \text{ for each } k = 1,\ldots,R \tag{4.4}$$

$$\sum_{j=1}^{N} y_j \le U \tag{4.5}$$

$$\sum_{j=1}^{N} I_j y_j \le B \tag{4.6}$$

Model (4.1) - (4.6) represents a fairly typical warehouse location formulation. The objective of this model is to determine the optimal warehouse network to deliver products from plants to customers such that the costs of transportation between plants and customers, variable warehouse handling costs, annual fixed warehouse operating costs and one-time warehouse location or construction costs are minimized. Constraints (4.2) assure that a model solution does not violate the capacity limitations of any of the M plants, while constraints (4.3) specify that a solution must satisfy the product demands of all R customer regions. Constraints (4.4) represent a customer service type requirement. They assure that each customer region receives all products from one warehouse (*i.e.*, a sole sourcing constraint). The parameter U in constraint (4.5) limits the number of warehouse locations selected and operated to a user specified maximum, while constraint (4.6) caps the maximum financial investment allowed in the warehouse network (*i.e.*, a budget constraint). Model (4.1)-(4.6) presents just one of the many variations of warehouse location models found in practice. Other common constraints found in warehouse location models but not shown in (4.1)-(4.6) include minimum and/or maximum capacity thresholds on individual warehouse locations. Note that in contrast to the integrated manufacturing and distribution models considered in Chapter 2, this model really only addresses the distribution side of the network. The location, the production output and the production mix of each network plant is fixed in advance for purposes of the warehouse location model. Then given these fixed manufacturing decisions, the warehouse location model attempts to determine the best warehouse network infrastructure to distribute product to customers. The efficacy of this approach must be evaluated on a situation by situation basis. Often for a myriad of reasons, this represents a very reasonable strategic planning approach for a firm that operates a manufacturing and distribution network. Further, for a pure distribution firm (e.g., a wholesaler), a warehouse location type model can address its network infrastructure requirements in a comprehensive, integrated fashion.

4.1.2 A Multi-Echelon, Multi-Transport Mode Distribution Problem

This section presents an integrated multi-echelon distribution and transportation planning problem commonly found in consumer products and other packaged goods industries. It has particular relevance to firms that serve a mix of small and large customers, some of whom order in full truckload or greater quantities, and some of whom order in relatively small quantities (e.g., less than truckload, LTL quantities). In reviewing the models in this section, we will explore several hierarchical dimensions including planning horizon and solution methodology.

Figure 4.4 depicts the distribution and transportation network assumed for the multi-echelon, multi-mode planning model (4.7) – (4.20) that will follow shortly. This particular problem assumes a distribution network consisting of major regional and/or national stocking warehouses (echelon 1) and local depot warehouses (echelon 2). For this example, we assume that the second echelon warehouses do not stock product but rather act as transshipment locations that may temporarily hold product for one or two days at the most. In practice, these second echelon warehouses may receive shipments from one or more regional or national stocking warehouses. These second echelon depots essentially then serve as consolidation points that aggregate smaller orders from different first echelon shipping locations all destined for the same ultimate customer receiving location. These second echelon depots may serve a single manufacturer/distributor shipping different products from multiple first echelon locations, or these second echelon depots may serve multiple manufacturers who all deliver to the same customers. The case of small local depots consolidating multiple manufacturers deliveries at transshipment points and then making "larger" consolidated deliveries to individual customers is often termed "pool distribution". Manufacturers who use pool distribution generally ship many small orders destined for the same general demand region on one large truckload to a pool distributor. The pool distributor receives these truckload shipments from multiple manufacturers, and then consolidates their collective smaller orders into one larger delivery to a specific local customer. Model (4.7) – (4.20) thus can represent the case of a two echelon network that exclusively serves a single manufacturer and/or distributor, or the case where the second echelon serves the distribution needs of multiple manufacturers and/or distributors.

Table 4.2 presents the notation for the multi-echelon, multi-modal distribution model (4.7) – (4.20).

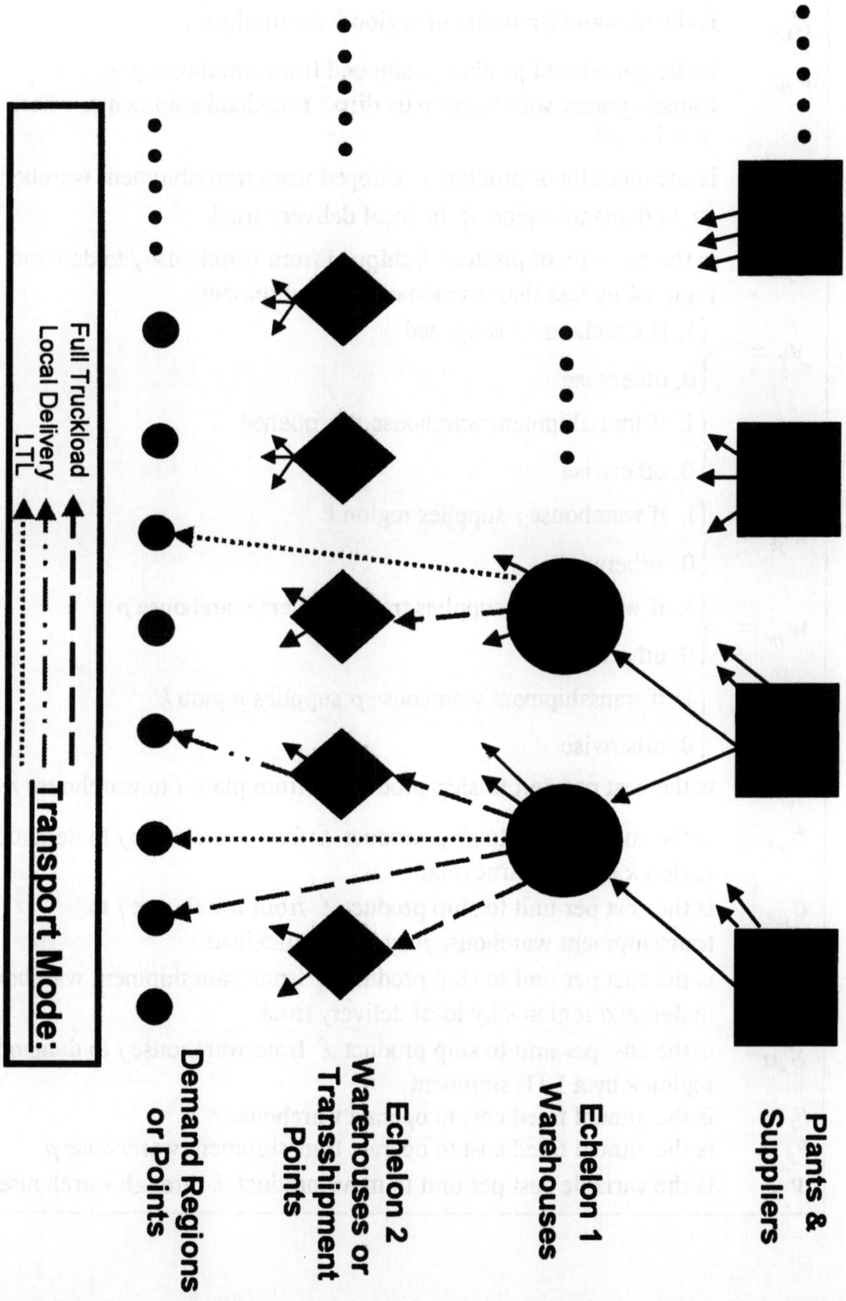

Figure 4.4 Multi-Echelon, Multi-Mode Distribution Network

Table 4.2 Notation For A Multi-Echelon, Multi-Mode Distribution Problem

$s_{ij\ell}$	is the quantity of product ℓ shipped from plant i to warehouse j, where $i = 1,\dots,M$; $j = 1,\dots,N$; and $\ell = 1,\dots,P$
$t_{jk\ell}$	is the quantity of product ℓ shipped from warehouse j to demand region k by direct truckload shipments, where $k = 1,\dots,R$
$d_{k\ell}$	is the demand (in units) of region k for product ℓ
$u_{jp\ell}$	is the quantity of product ℓ shipped from warehouse j to transshipment warehouse p by direct truckload shipments, where $p = 1,\dots,O$
$x_{pk\ell}$	is the quantity of product ℓ shipped from transshipment warehouse p to demand region k by local delivery truck
$z_{jk\ell}$	is the quantity of product ℓ shipped from warehouse j to demand region k by less than truckload (LTL) shipments
$y_j =$	$\begin{cases} 1, & \text{if warehouse } j \text{ is opened} \\ 0, & \text{otherwise} \end{cases}$
$y_p =$	$\begin{cases} 1, & \text{if transshipment warehouse } p \text{ is opened} \\ 0, & \text{otherwise} \end{cases}$
$w_{jk} =$	$\begin{cases} 1, & \text{if warehouse } j \text{ supplies region } k \\ 0, & \text{otherwise} \end{cases}$
$w_{jp} =$	$\begin{cases} 1, & \text{if warehouse } j \text{ supplies transshipment warehouse } p \\ 0, & \text{otherwise} \end{cases}$
$w_{pk} =$	$\begin{cases} 1, & \text{if transshipment warehouse } p \text{ supplies region } k \\ 0, & \text{otherwise} \end{cases}$
$a_{ij\ell}$	is the cost per unit to ship product ℓ from plant i to warehouse j
$b_{jk\ell}$	is the cost per unit to ship product ℓ from warehouse j to demand region k by direct truckload
$c_{jp\ell}$	is the cost per unit to ship product ℓ from warehouse j to transshipment warehouse p by direct truckload
$e_{pk\ell}$	is the cost per unit to ship product ℓ from transshipment warehouse p to demand region k by local delivery truck
$g_{jk\ell}$	is the cost per unit to ship product ℓ from warehouse j to demand region k by a LTL shipment
F_j	is the annual fixed cost to operate warehouse j
F_p	is the annual fixed cost to operate transshipment warehouse p
$v_{j\ell}$	is the variable cost per unit to move product ℓ through warehouse j

Table 4.2 Notation For A Multi-Echelon, Multi-Mode Distribution Problem (continued)

$v_{p\ell}$	is the variable cost per unit to move product ℓ through transshipment warehouse p
I_j	is the one-time investment cost required to build or open a new warehouse j
I_p	is the one-time investment cost required to build or open a new transshipment warehouse p
$C_{i\ell}$	is the annual production capacity of plant i to manufacture product ℓ
U_j	is the maximum number of (stocking) warehouses that the firm wants to open or operate
U_p	is the maximum number of transshipment warehouses that the firm wants to open or operate
B^N	is the maximum financial investment that the firm wants to make in opening new (stocking) warehouses
B^0	is the maximum financial investment that the firm wants to make in opening new transshipment warehouses
B	is the maximum financial investment that the firm wants to make in opening new (stocking) and transshipment warehouses
T_j	is the annual throughput capacity (in units) of warehouse j
T_p	is the annual throughput capacity (in units) of transshipment warehouse p
Q_ℓ	is the weight (*i.e.*, lbs) per unit of product ℓ
MW_{jp}	is the minimum total lbs. per period (e.g., per year) that must flow between warehouse j and transshipment warehouse p in order for warehouse j to be allowed to supply transshipment warehouse p
MW_{jk}	is the minimum total lbs. per period that must flow between warehouse j and demand region k in order for warehouse j to be allowed to supply demand region k by direct truckload on a regular basis (*i.e.*, to use direct truckloads as the standard replenishment mode)

A Strategic Multi-Echelon Multi-Mode Distribution Problem

Minimize:

$$\sum_{i=1}^{M}\sum_{j=1}^{N}\sum_{\ell=1}^{P} a_{ij\ell}s_{ij\ell} + \sum_{j=1}^{N}\sum_{k=1}^{R}\sum_{\ell=1}^{P}\left(b_{jk\ell}t_{jk\ell} + g_{jk\ell}z_{jk\ell} + w_{jk}v_{j\ell}d_{\ell k}\right)$$

$$+\sum_{j=1}^{N}\sum_{p=1}^{O}\sum_{\ell=1}^{P} c_{jp\ell}u_{jp\ell} +$$

$$\sum_{p=1}^{O}\sum_{k=1}^{R}\sum_{\ell=1}^{P}\left(e_{pk\ell}x_{pk\ell} + w_{pk}v_{p\ell}d_{\ell k}\right) + \sum_{j=1}^{N}\left(y_jF_j + y_jI_j\right)$$

$$+\sum_{p=1}^{O}\left(y_pF_p + y_pI_p\right) \qquad (4.7)$$

Subject to:

$$\sum_{j=1}^{N} s_{ij\ell} \le C_{i\ell} \text{ for all } i = 1,...,M \text{ and } \ell = 1,...,P \tag{4.8}$$

$$\sum_{i=1}^{M}\sum_{\ell=1}^{P} s_{ij\ell} \le T_j \text{ for all } j = 1,...,N \tag{4.9}$$

$$\sum_{j=1}^{N}\sum_{\ell=1}^{P} u_{j\ell} \le T_p \text{ for all } p = 1,...,O \tag{4.10}$$

$$\sum_{j=1}^{N}\left(t_{jk\ell} + z_{ik\ell}\right) + \sum_{p=1}^{O} x_{pk\ell} - d_{k\ell} = 0$$

$$\text{for all } k = 1,...,R \text{ and } \ell = 1,...,P \tag{4.11}$$

$$\sum_{j=1}^{N} w_{jk} = 1 \text{ for all } k = 1,...,R \tag{4.12}$$

$$\sum_{p=1}^{O} w_{pk} = 1 \text{ for all } k = 1,...,R \tag{4.13}$$

$$\sum_{j=1}^{N} y_j \le U_j \tag{4.14}$$

$$\sum_{p=1}^{O} y_p \le U_p \tag{4.15}$$

$$\sum_{j=1}^{N} I_j y_j \le B_j \tag{4.16}$$

$$\sum_{p=1}^{R} I_p y_p \le B_p \tag{4.17}$$

$$\sum_{j=1}^{N} I_j y_j + \sum_{p=1}^{R} I_p y_p \le B \tag{4.18}$$

$$\sum_{\ell=1}^{P} w_{jp} u_{jp\ell} Q_\ell \ge MW_{jp} \text{ for all } j = 1,...,N \text{ and } p = 1,...,O \tag{4.19}$$

$$\sum_{\ell=1}^{P} w_{jk} t_{jk\ell} Q_\ell \ge MW_{jk} \text{ for all } j = 1,...,N \text{ and } k = 1,...,R \tag{4.20}$$

Model (4.7) – (4.20) can provide both strategic and tactical planning decision support for a network with multiple distribution echelons and multiple transportation modes to serve those echelons. This model develops a plan to meet customers' product demands while minimizing the sum of transportation costs, one-time warehouse location costs, and the variable and annual fixed costs of warehouse operations. Note that three modes of transportation are available to deliver products to customers including: (1) direct truckload shipments from echelon 1 warehouses, (2) direct less than truckload (LTL) shipments from echelon 1 warehouses, and (3) local deliveries from transshipment warehouses (echelon 2). In its present form, one would classify (4.7) – (4.20) as a strategic planning model in the sense that it evaluates what is typically a multi-year decision and investment, namely, the location of major stocking warehouses (echelon 1). We will observe shortly that with minor modifications, this model also can serve as a tactical/annual decision support planning tool.

Constraints (4.8) assure that a solution to this model does not violate the production capacity constraints of the supply plants of the network. Equations (4.9) and (4.10) constrain the model from violating the respective throughput capacity constraints of the stocking warehouses (echelon 1) and of the transshipment warehouses (echelon 2), respectively. Constraints (4.11) assure that a model solution satisfies the respective demands of the R customer regions for the P products. Equations (4.12) and (4.13) represent single-sourcing, customer service constraints. Specifically, constraints (4.12) stipulate that all direct shipments from an echelon 1 warehouse j to a demand region k must originate from the same warehouse (*i.e.*, a single echelon 1 source of supply for each customer). Similarly, constraints (4.13) specify that a maximum of one transshipment warehouse p (echelon 2) can serve each demand region k. Constraints (4.14) and (4.15) place an upper bound on the respective number of echelon 1 and echelon 2 warehouses that a firm will consider opening and operating on its network. Equations (4.16) perform a similar bounding role by setting upper limits on the maximum financial investment that a firm will make in opening echelon 1 warehouses, while equations (4.17) place a limit on the financial investment allowed in opening echelon 2 warehouses. Constraints (4.18) bound the total financial investment permitted for echelons 1 and 2 in total. Note that in practice, a firm may choose to use only a subset of these three financial constraints (e.g., only 4.18). Further, the parameters I_j, B_j, I_p, B_p and B are sometimes expressed as amortized annual cost values rather than as the full one-time costs of opening a warehouse, depending upon the modeling application. Equations (4.19) and (4.20) represent a combination of what could be termed implicit customer service and operating efficiency constraints. Constraints (4.19) dictate that a stocking warehouse j cannot supply a transshipment warehouse p unless the total lbs shipped from j to p during the planning horizon (e.g., one year) exceeds a defined minimum total lbs (MW_{jp}). For example, assume that whenever a warehouse j ships a truckload to a transshipment warehouse p, it must ship a minimum of 40,000 lbs. Under such a policy, a value of 6,240,000 lbs. for a particular MW_{jp} over an annual planning horizon would mean that in order for a warehouse j to be designated a replenishment supply location for a transshipment warehouse p, warehouse j would have to make at least one hundred and fifty-six 40,000 lb. shipments per year (or an average of three 40,000 lb shipments per week) to warehouse p. Thus, one can argue that (4.19) implicitly represents a customer service constraint because it essentially assures some desired minimum shipping frequency must exist (or be achievable) between a stocking warehouse j and a transshipment warehouse p in order for j to serve as a normal replenishment location for p. This implicit minimum frequency constraint helps to assure a reasonable order cycle and delivery time between echelon 1 and echelon 2 warehouses. This in turn facilitates reasonable order cycle times between an echelon 1 stocking warehouse and the ultimate customer k. Constraint (4.20) plays a similar role to assure reasonable shipment frequencies between echelon 1 distribution centers and customers receiving direct shipments (from echelon 1) via truckload. Specifically, (4.20) dictates that in order for a warehouse j to be designated the regular planned replenishment location for a demand region k, a minimum total lbs. (MW_{jk}) must flow annually (via truckload) between the warehouse and demand region. This again assures some minimum shipping frequency and some minimum order cycle time.

(4.7) - (4.20) represents a strategic planning model that will provide decision support for establishing and/or modifying a distribution infrastructure and supply plan to deliver products from manufacturing locations through to end customers. In this example, the model will recommend the locations for both echelon 1 and echelon 2 distribution facilities, as well as the transport modes to employ for moving products between locations, and the specific supply patterns to employ (*i.e.*, which facilities should supply which products to other facilities and customers). We will discuss the mode choice selection aspect of this model shortly during a review of solution approaches for this model. However, first we consider how this model can serve in different planning roles and also how it fits into a hierarchical planning structure.

First of all, even for strategic planning purposes, it is worth noting that in practice many firms and planners will utilize only a subset of the constraints (4.8) - (4.20), or they will employ additional constraints not described in this formulation. For example, practitioners frequently will not explicitly employ budget constraints (4.16 - 4.18) or minimum volume flow constraints (4.19 - 4.20). Additionally, minimum volume or capacity constraints are also frequently placed on potential locations. (This example model places only maximum capacity constraints [4.9 - 4.10] on the distribution facilities.)

This multi-mode, multi-echelon model can also provide ongoing tactical/annual planning decision support with minor modifications to the initial formulation. Assume that a firm already has its echelon 1 warehouses in place and that has it no plans to change either the number, the capacity or the locations of these echelon 1 facilities. A planner can use a modified version of this model to address a number of ongoing tactical planning issues including:

- revising the replenishment supply patterns in response to either capacity imbalances between individual echelon 1 warehouses or changes in costs,
- revising mode choices for particular transport lanes, and
- changing the number, locations and/or supply responsibilities of echelon 2 warehouses

In particular, if the echelon 2 warehouses on a network represent small third party depots, a firm often has considerable flexibility to start up and/or shut down use of such facilities within a typical tactical planning horizon of 12 to 18 months. Model (4.21) which represents a simplified version of the strategic multi-mode, multi-echelon formulation, can address the tactical issues just described. Note that the last term in the objective function ($y_p I_p$) can be omitted if there is no one time investment cost associated with starting up a transshipment warehouse. Conversely, if there are one time costs associated with shutting down a transshipment warehouse, these should be added to the model formulation.[3]

This model provides the necessary decision support to construct and update tactical infrastructure and supply replenishment plans. What remains to establish now consists of the linkage between this tactical/annual distribution and transportation mode planning tool and the related shorter run activities that occur in distribution and transportation planning and scheduling.

A Tactical Multi-Echelon Multi-Mode Distribution Problem

Minimize:

$$\sum_{i=1}^{M}\sum_{j=1}^{N}\sum_{\ell=1}^{P} a_{ij\ell} s_{ij\ell}$$

$$+\sum_{j=1}^{N}\sum_{k=1}^{R}\sum_{\ell=1}^{P}\left(b_{jk\ell} t_{jk\ell} + g_{jk\ell} z_{jk\ell} + w_{jk} v_{j\ell} d_{\ell k}\right)$$

$$+\sum_{j=1}^{N}\sum_{p=1}^{O}\sum_{\ell=1}^{P} c_{jp\ell} u_{jp\ell}$$

$$+\sum_{p=1}^{O}\sum_{k=1}^{R}\sum_{\ell=1}^{P}\left(e_{pk\ell} x_{pk\ell} + w_{pk} v_{p\ell} d_{\ell k}\right)$$

$$+\sum_{p=1}^{O}\left(y_p F_p + y_p I_p\right) \tag{4.21}$$

Subject to: 4.8 - 4.13, 4.15, 4.19, and 4.20

For illustrative purposes, assume that using models (4.7) - (4.20) and (4.21), as well as other quantitative and qualitative inputs, a firm has established and now operates a two echelon distribution network with a mix of transportation modes. Further, assume that the firm employs one of the numerous commercial software packages available to schedule pickups and deliveries by the carriers and modes that serve the firm's network.[4] The "top-down" hierarchical relationship that exists in this example is simply that the firm's strategic and then tactical distribution and transportation planning activities dictate what product flows occur

[3] Particularly in the case of small regional third party pool operations, there often is no one-time cost of closing a third party relationship. Rather, there may simply be a contractual obligation to provide some minimum notice period of intent to end the contractual relationship (e.g., 60 days).

[4] This example applies regardless of how a firm schedules its transportation pickups and deliveries (e.g., manually, with customized scheduling software, commercial package software, etc.)

at the operational level between particular origin-destination pairs on the network. That is, the plan set at the strategic and tactical levels guides the day-to-day transportation deliveries that operational planners schedule. A critical element necessary to assure success at the operational, executional level consists of the "bottom-up" feedback that this level must provide to the tactical planning level. Table 4.3 offers an example of the type of feedback that must flow "up" the transportation/distribution planning hierarchy.

Table 4.3 Illustrative Operational Monitoring And Feedback System

(1) Transshipment Point Or Local Warehouse (p=1,...,O)	(2) Number of Shipments From Regional Warehouse To Local Warehouse During Last 10 Weeks	(3) Average Shipments Per Week From Regional Warehouse To Local Warehouse		(4) Average Lbs. Per Shipment		(5) Average Freight Cost Per Lb.	
		Planned	Actual	Planned	Actual	Planned	Actual
1	55	5.0	5.5	36,000	37,000	$.04	$.04
2	50	5.0	5.0	35,000	38,000	$.03	$.03
3	30	3.0	3.0	37,000	40,000	$.04	$.04
4	20	3.0	2.0	35,000	17,000	$.04	$.08
5	45	4.0	4.5	35,000	36,000	$.03	$.03
• • • • • •	• • • • • •	• • • • • •	• • • • • •	• • • • • •	• • • • • •	• • • • • •	• • • • • •
Total/ Average							

Specifically, Table 4.3 presents an illustrative chart that monitors the shipment flow between the echelon 1 stocking warehouses and the echelon 2 transshipment warehouses. Column 2 displays the total shipments each transshipment warehouse has received over some recent period of time (e.g., 10 weeks), while column 3 compares the planned and actual average shipments per week received by each warehouse. Note that transshipment warehouse 4 has received an average of only 2 shipments per week rather than the planned 3 per week over the last 10 weeks. Further, column 4 shows that the actual average weight (17,000 lbs.) per shipment to this warehouse has slipped well below the planned target (35,000 lbs.); and consequently, as shown in column 5, the actual average transportation cost per lb. ($.08) delivered to this warehouse has exceeded the planned cost ($.04 per lb.). This monitoring chart indicates that the distribution and transportation plans developed at the tactical level for transshipment warehouse 4 have not materialized at the operational level. This suggests that the firm must reconsider the viability of the warehouse 4 operation. In order to do this, planners at the tactical level must use the most current demand history and forecasts, model (4.21) and other inputs to evaluate alternative options for improving the frequency of service to the region(s) served by warehouse 4, and at the same time for reducing the transportation costs to serve this region(s). It may turn out that insufficient demand exists to maintain the transshipment warehouse 4 operation, or that warehouse 4 must serve a broader geographic region to generate sufficient total throughput volume. This simple example illustrates the type of feedback that must flow from the operational planning level back to the tactical planning level on a regular basis.

Hierarchical Solution Approaches To The Multi-Mode, Multi-Echelon Distribution Problem

For applied modeling purposes, there exist many different approaches to solving problems such as (4.7) - (4.20) and (4.21). Some of these practical approaches are not necessarily documented in textbooks or articles. However, most methodologies typically involve a decomposition approach, whether it be an elegant mathematical decomposition methodology or an extemporaneous real world approach to somehow restructure the full problem into a smaller problem or series of linked problems. Most solution methodologies also involve aggregating the true level of problem detail to a level that somehow reduces the overall problem size without significantly, if at all, denigrating the solution. In this section, we explore a few common "practical" approaches to problems (4.7) - (4.20) and (4.21).

A first standard approach consists of aggregating the demand locations in the real world problem to some "aggregated" set of demand points that reflect the collective demands of these individual locations. This represents such a common approach that in some respects it virtually requires no discussion other than to note its importance in developing practical planning models. In the context of the multi-mode, multi-echelon distribution problem, however, this aggregation of demand also becomes part of a classic decomposition approach to this problem. A discussion of constraints (4.19) and (4.20) serves to illustrate the principles of "applied" aggregation and decomposition techniques used for this sort of problem.

As previously described, constraints (4.20) assure that a stocking warehouse in echelon 1 can only serve a demand region k with direct truckload service if that region k can generate some defined minimum total lbs. of demand over the planning horizon. This constraint implicitly guarantees that only regions that demand a sufficient volume to guarantee acceptable direct truckload shipment frequencies and lbs. per truckload will receive truckload service. Because full truckload service generally represents a point-to-point move (e.g., a firm's outbound shipping warehouse to a customer's receiving warehouse), one would really have to define model demand regions as individual customer receiving locations in order to accurately specify this constraint. Many firms ship to thousands of customer locations, and to formulate a model with thousands of individual demand locations often does not represent a viable option from either a mathematical solution point of view, and/or a usable planning output perspective. For this reason, it is often necessary to define a group of individual customer receiving locations in close geographic proximity to each other as one demand region. This need to aggregate demand points ties in very well with a common real world approach to modeling constraints (4.20) and (4.19). Specifically, the following methods often work quite well as surrogates to explicitly modeling constraints (4.20) and (4.19). For illustrative purposes, we focus on constraints (4.20) for the moment.

Method 1

Step 1 - Define the set R of k demand regions for model (4.7) - (4.20). There exist countless informal and formal techniques for creating such regions. For example, one method of defining demand regions for a model of the United States consists of creating aggregations of contiguous three digit postal codes centered around Metropolitan Statistical Areas (MSA's). In an applied study, one can usually obtain an adequate level of accuracy by modeling the U.S. with between 100 and 150 demand regions (see House and Jamie, 1981).

Step 2 - Define two categories or classes of demand: (A) full truckload demand and (B) less than full truckload demand. Direct truckload shipments from echelon 1 stocking warehouses to individual customer receiving locations will serve full truckload (*i.e.*, class A) demand; while either pool shipments moving from echelon 1 warehouses through transshipment echelon 2 warehouses to individual customer receiving locations, or direct LTL shipments from echelon 1 warehouses to individual customer locations will serve less than full truckload (*i.e.*, class B) demand.

Step 3 - Create the initial demand data for model (4.7) - (4.20) using recent historical order and shipment data (e.g., from the most recent 12 months). Based upon what transportation mode (*i.e.*, truckload, LTL or pool) was actually used to ship each individual order, classify all historical orders/shipments as either class A (*i.e.*, full truckload) demand, or class B (*i.e.*, LTL and pool) demand.

Step 4 - Aggregate into one demand data point all the class A individual shipments during the recent historical period shipped into a defined demand region k (*i.e.*, create an aggregate demand value). Perform this same aggregation process for each of the R demand regions defined in step 1.

Step 5 - Aggregate into one demand data point all the class B individual shipments during the recent historical period shipped into a defined demand region k (*i.e.*, create an aggregate demand value). Perform this same aggregation process for each of the R demand regions defined in step 1.

Step 6 - Stop. The initial demand history for the model has now been created. A planner can now make forecast adjustments or other modifications to this demand history as appropriate to serve the particular modeling decision support application.

Method 2

This method uses the same steps as method 1 except for step 3. Method 2 utilizes the following alternative to step 3 in method 1.

1. Evaluate the total volume (lbs.) shipped to each individual customer receiving location over a recent historical period, or alternatively the projected total shipment volume over a defined future planning horizon.
2. Determine whether the total lbs. that a customer location did (or will) receive over the historical (or future) period exceeds the minimum weight threshold required for full truckload service. (E.g., If 40,000 lbs. per week represents the minimum weight requirement for full truckload service, then a customer location must have received over 2,080,000 lbs. during the most recent 52 weeks of history.)
3. Based on the preceding calculations (in 2), classify each customer receiving location as either a class A (full truckload) demand location if it meets the minimum weight threshold, or as a class B demand location if it does not meet the full truckload weight threshold.

Methods 1 and 2 both eliminate the need to maintain constraints (4.20) in the multi-echelon, multi-mode distribution model. They accomplish this by using a pre-processing approach to determine in advance of the actual optimization modeling process which transport modes will serve which individual customer locations. This pre-processing or staged approach consists of first separating total demand into sub-classes of "transport mode related" demand. This hierarchical or decomposition approach thus allows a planner to eliminate constraints (4.20) from the original planning model. Method 1 implicitly assumes that the historical mix of mode shipments to individual customer locations represents a good predictor of an efficient future mode mix to these same locations. In contrast, method 2, which evaluates individual customer receiving locations on the basis of past or projected total weight (*i.e.*, demand), does not assume that the previous mix of transport modes represents a good predictor of an efficient future mode mix. In this respect, method 2 often offers additional analytic flexibility relative to method 1. Regardless of whether a planner uses either of these two methods, or some other variation upon these two, this multi-stage modeling approach reduces the overall complexity of the original problem.

One can eliminate constraints (4.19) using similar decomposition constructs to those just illustrated for constraint (4.20). The additional complexity posed by the multiple end customer locations served by the truckload shipments from echelon 1 warehouses j to echelon 2 warehouses p typically requires slightly more detailed decomposition approaches. Nevertheless, using similar techniques to those illustrated in methods 1 and 2, a planner can also eliminate the need for constraints (4.19) if desired.

This completes our discussion of practical hierarchical decomposition methods to solve the multi-echelon, multi-mode distribution problem reviewed in this section. In practice, there exist a wide array of methods for treating complex problems such as this one in a staged approach. The methods just reviewed provide a few simple examples of the types of decomposition methodologies that form an important part of applied transportation and distribution modeling. Further, as illustrated in Table 4.3, the successful use of such models to support integrated distribution and transportation planning requires a series of linkages between the strategic, tactical and operational planning and scheduling of these activities.

4.1.3 Strategic Rail-Based Distribution And Transportation Planning

This section examines another example of strategic distribution and transportation infrastructure and mode choice planning. In this example, we review a project conducted by a leading international automobile manufacturer to determine the best transport mode and rail network location strategy to supply its North American dealers. This case study provides another illustration of the type of strategic distribution planning that typically occurs independently from manufacturing strategic planning (*i.e.*, in a non-integrated fashion). This case also serves to provide a perspective on rail and intermodal distribution planning type activities.

Case Study: Transport Network Design And Mode Choice Modeling For Automobile Distribution[5]

In the early 1990's, a leading international auto manufacturer wished to establish a plan for its future vehicle distribution that would determine the best mix of truck, conventional rail and containerized intermodal rail to transport cars and light trucks from its North American plants and ports of entry to its dealers. This firm enlisted the services of a consulting firm to assist in this strategic evaluation and the two parties decided to develop a mixed integer programming model to assist in determining the manufacturer's best transport mode and rail network location strategy. The transportation of automobiles from the assembly plant to the dealer involves one or more transportation modes: truck, rail, and/or ship and is frequently intermodal as it utilizes more than one of these modes. The overriding questions of interest to the manufacturer as it embarked upon this study included:

1) what would be the best modal mix of truck, conventional rail and containerized intermodal rail to transport cars and light trucks from plants and ports of entry to its dealers? and,

2) what would be the optimal number, types and locations of rail terminals required for its distribution network in the future?

Company And Automobile Freight Transport Industry Background

Motor vehicle manufacturers were selling at a rate of over 15 million vehicles in North America around 1994. Vehicles were shipped from over 100 origin points (assembly plants and ports of entry) to over 30,000 destination locations (dealers

---

[5] The reader interested in the full details of this case study is referred to Miller, Wise and Clair (1996) for further reading. This section draws heavily from this article.

and large fleet locations). This extensive distribution system resulted in a wide variety of geographically convenient choices for consumers, but also meant that most dealers received less than one vehicle per day of any given model. Rather than ship single vehicles from each plant to each dealer, the North American vehicle distribution system utilized a network of about 150 rail-truck terminals, typically based in large metropolitan areas, to take advantage of the long distance, high volume efficiencies of railroads and the short distance flexibility of motor carriers.

At the plants and ports, about 70 percent of the vehicles were loaded onto bilevel or trilevel railcars, 8-18 vehicles per railcar, and then shipped to a destination rail terminal. At these terminals, the railcars were unloaded and the vehicles were sorted into bays for specific dealers. Haulaway motor carriers would then load 8-10 vehicles onto their rigs at these terminals, and deliver their load to 2-3 dealers in the metropolitan area. Haulaway motor carriers also delivered about 30% of all vehicles from the manufacturer's plants and ports of entry directly to dealers.

By the mid-1990's, railroads had invested heavily in fully enclosed railcars to reduce damage in transit. In fact, manufacturers' interest in "zero defects" led some to go even further and explore transportation modes that would be completely damage-free. One major manufacturer, the subject of this case, was quite interested in using containers as a means to achieve damage free transit. Through a cleverly designed steel rack and loading system, a transportation equipment supplier had developed a method to load up to six automobiles into a 48-foot container, providing a completely enclosed transportation mode. The vehicle manufacturer who is the subject of this case introduced this "vehicle containerization system" into several traffic lanes in the early 1990s and saw promise in further expansion across its network. This was the initial problem posed for network modeling: given this new containerized mode of transportation, where should it be introduced (*i.e.*, displacing either the existing direct motor carrier haulaway mode or rail multilevel/haulaway mode) and how would it affect overall network performance (total cost and average speed from the time the vehicle was released from a plant until the time it was delivered to a dealer)?

Costs

Figure 4.5 defines the key cost components which the network model had to evaluate. These costs represented all of the major logistics costs of delivering vehicles from a plant or port of entry to the dealer.

Transport Modes

As is frequently the case in applied studies, numerous versions of a general model were developed to explore a wide range of transport scenarios. In the most general case, however, four major alternative "transport modes or methods" by which the manufacturer could ship a vehicle from its origin (a plant or port) to a dealer were modeled. These four alternatives included:

Plant Loading :	The cost to load a vehicle onto a rail car or a haulaway truck at a plant or port once the vehicle leaves the production line or ship.
Container Drays:	The cost per vehicle to transport an intermodal container of vehicles by truck. This activity could occur at several different points in the delivery process.
Rail Linehaul:	The cost to ship a vehicle by rail between an origin and destination rail ramp.
Haulaway Linehaul:	The cost to ship a vehicle by haulaway truck between an origin and a destination.
Inventory Carrying:	The daily inventory carrying cost for each vehicle type based on the manufacturer's defined daily model capital carrying cost.
Terminal Loading:	The cost to load a vehicle onto a rail car at a rail terminal.
Terminal Unloading:	The cost to unload a vehicle from a rail car at a rail terminal.
Damage:	The cost of the damage to a vehicle which occurs in transit between the origin plant or port and the destination dealer. (*I.e.*, The manufacturer quantified the costs per vehicle associated with paint nicks, dirt, etc.. incurred in transit.)

Figure 4.5 Description Of Logistics Costs Modeled

1) *Haulaway direct*: vehicles are shipped directly from a plant to a dealer by a haulaway truck
2) *Multilevel rail*: vehicles are shipped by conventional multilevel rail car from a rail yard at or nearby a plant to a destination rail yard, and then delivered to the dealer on a haulaway truck
3) *Intermodal rail*: vehicles are shipped in containers on double-stack rail cars from an intermodal rail yard nearby a plant to a destination intermodal rail yard and/or rail auto terminal, and then delivered by haulaway truck
4) *Multilevel rail via Mixing Centers*: vehicles are shipped by conventional multilevel rail car from a plant to an intermediate rail yard containing a Mixing Center. Vehicles are then sorted, reloaded and shipped from the

intermediate rail yard to a destination rail yard. Delivery to the dealer is on a haulaway truck.

Haulaway direct (alternative 1) represents the simplest activity chain of the four alternatives. Vehicles are simply loaded onto the haulaway truck and transported directly to a dealer. A number of individual activities jointly compose the "multilevel rail" option (alternative 2). While the rail linehaul typically represents the major activity, there is also at least one haulaway truck move from the destination rail yard to the dealer. Additionally, depending upon whether or not a plant or port has a rail yard adjacent to it, there is often a haulaway truck move required to transport a vehicle from the plant/port to the origin rail yard. The intermodal option (alternative 3) is perhaps the most complex from a process perspective. In addition to the double-stack rail linehaul, this activity chain can include the following:

- haulaway truck moves from the plant/port to an origin autoramp and from a destination autoramp to the dealer, and
- container drays from the origin autoramp to an origin intermodal rail yard, and from a destination intermodal rail yard to a destination autoramp.

Transport alternative 4 (multilevel rail via a mixing center) in reality represents a variation on option 2, rather than a fourth modal alternative. A mixing center is an intermediate location where vehicles originating from different plants/ports are matched up with other vehicles destined for the same final rail yard. After this mixing and matching activity occurs at the mixing center, trains depart transporting these "mixed" vehicles via multilevel rail linehaul to their destination rail yard. Briefly, by shipping its vehicles to a mixing center rather than directly to a destination rail yard, a plant could improve the velocity with which it dispatches vehicles from a plant once the vehicles exit the production line. In conventional multilevel rail (option 2), a plant must wait until it accumulates a minimum of a rail car of vehicles all destined for the same final rail yard before it can ship vehicles. Under the mixing center alternative, the plant can ship a rail car of vehicles to a mixing center as soon as it has produced enough vehicles for a much larger destination region (e.g., the "west coast of the United States" instead of a single metropolitan area) to fill a rail car, regardless of the ultimate rail yard destination of each vehicle. Note that the additional activities composing transport alternative 4 are identical to those activities which occur in option 2.

The breadth of the vehicle manufacturer's product lines necessitated that approximately thirty-five product families representing over twenty major vehicle types or carlines be defined. Each vehicle type itself represented up to twenty or so unique vehicles which differed by color and/or minor model gradations. The need to define more product families than carlines resulted from the vehicle manufacturers' desire to maintain product manufacturing and sourcing as fixed throughout the modeling exercise. To accommodate this restriction, in cases where multiple plants could produce the same major vehicle type, multiple product families were defined - one family for each facility.

The manufacturer's U.S. and Canadian dealership network totaled over 6,000 individual dealers, and approximately 275 dealer regions were created to represent these individual dealers in the model. The dealer regions were created by aggregating the annual demands of all dealers within defined contiguous three digit zip code areas. As previously discussed, in an applied study, one can usually obtain an adequate level of accuracy by modeling the U.S. and/or Canada with between 100 and 150 demand regions (House and Jamie, 1981). However, given this study's focus on modal transport alternatives, and the sensitivity of the relative competitiveness of these alternatives to the total distance between a plant and a dealer; it was decided to create relatively "tight" demand regions.

To model all four modal choices, about 170 facilities were defined. This included approximately 25 plants and ports, and about 145 rail yards. The rail terminals consisted of a mix of conventional multilevel vehicle loading and unloading ramps, intermodal facilities and mixing centers. Table 4.4 provides the notation for the model formulation, and the actual model developed follows this table.

Table 4.4 Notation For The Mode Choice And Rail Terminal Location Model

j	is the index for vehicle product families, j=1,2,..., J
p	is the index for plants and ports of entry, p = 1,2,..., PP. (For purpose of brevity, in all following definitions we will omit "ports," and plants will refer to both plants and ports)
d	is the index for automobile and light truck (vehicle) dealers, d = 1,2,...,DD
r	is the index for destination multilevel rail yards, r = 1,2,..., RR
i	is the index for destination intermodal terminals, i = 1,2,..., IM
m	is the index for multilevel rail mixing centers, m = 1,2,..., MC
O_{pjd}	is the number of vehicles of family j shipped directly to dealer d from plant p by haulaway truck
R_{pjr}	is the number of vehicles of family j shipped to destination rail yard r from plant p by multilevel rail
T_{rjd}	is the number of vehicles of family j shipped to dealer d from destination rail yard r by haulaway truck
I_{pji}	is the number of vehicles of family j shipped to destination intermodal yard i from plant p by intermodal transport
S_{ijd}	is the number of vehicles of family j shipped to dealer d from destination intermodel yard i by haulaway truck
W_{pjm}	is the number of vehicles of family j shipped to multilevel rail mixing center m from plant p
X_{mjr}	is the number of vehicles of family j shipped to destination rail yard r from rail mixing center m
A_{pjd}	is the unit cost to ship one vehicle of family j directly to dealer d from plant p by haulaway truck
B_{pjr}	is the unit cost to ship one vehicle of family j to destination rail yard r from plant p by multilevel rail
C_{rjd}	is the unit cost to ship one vehicle of family j to dealer d from destination rail yard r by haulaway truck
D_{pji}	is the unit cost to ship one vehicle of family j to destination intermodal yard i from plant p by intermodal transport

Table 4.4 Notation For The Mode Choice And Rail Terminal Location Model (continued)

E_{ijd}	is the unit cost to ship one vehicle of family *j* to dealer *d* from destination intermodal yard *i* by haulaway truck
F_{pjm}	is the unit cost to ship one vehicle of family *j* to multilevel rail mixing center *m* from plant *p*
G_{mjr}	is the unit cost to ship one vehicle of family *j* to destination rail yard *r* from rail mixing center *m*
H_j	is the daily inventory carrying cost of one vehicle of family *j*
Z_{oe}	is the transit time (in days) from an origin *o* to a destination *e*; where *o* is a placeholder for *p, r, i* and *m*; and where *e* is a placeholder for *d, r, i* and *m*- for all valid combinations of *o* and *e*
K_{pje}	is the unit cost to ship (by haulaway truck) one vehicle of family *j* from plant *p* to the origin autoramp/rail yard associated with plant *p* and destination *e*; where *e* is a placeholder for *r, i* and *m*. (Note that each plant *p* uses one assigned origin autoramp/rail yard for any shipment to a particular destination *e*)
Q_{pji}	is the unit cost per vehicle of family *j* of the origin and destination container dray moves required on shipments between plant *p* and destination intermodal yard *i*
M_{oje}	is the cost of the damage to a vehicle of family *j* which occurs in transit between an origin *o* and destination *e*; where *o* is a placeholder for *p, r, i* and *m;* and where *e* is a placeholder for *d, r, i* and *m*
L_{oje}	is the cost to load a vehicle of family *j* onto a haulaway truck, rail car or container at an origin *o* (e.g. from the production line of plant *p*); where 0 is a placeholder for *p, r, i* and *m;* and where *e* is a placeholder for *d, r, i* and *m*
U_{je}	is the cost to unload a vehicle of family *j* at a destination *e*; where *e* is a placeholder for *d, r, i* and *m*
V_{mj}	is the cost of all activities other than loading and unloading associated with moving a vehicle of family *j* through a rail mixing center *m*
N_{jd}	is the demand for vehicle family *j* at dealer *d*
MV_m	denotes the minimum annual volume of vehicles which must pass through mixing center *m*, in order for *m* to operate
MV_i	denotes the minimum annual volume of vehicles which must pass through intermodal yard *i*, in order for *i* to operate
Y_m	= 1 if a rail mixing center is opened, 0 otherwise
P_m	is the amortized annual fixed cost to build a new mixing center *m*

The Mode Choice And Rail Terminal Location Model

Minimize:

$$\sum_p\sum_j\sum_d O_{pjd}\left(A_{pjd}+H_jZ_{pd}+M_{pjd}+L_{pjd}+U_{jd}\right)$$
$$+\sum_p\sum_j\sum_r R_{pjr}\left(B_{pjr}+H_jZ_{pr}+M_{pjr}+L_{pjr}+U_{jr}+K_{pjr}\right)$$
$$+\sum_r\sum_j\sum_d T_{rjd}\left(C_{rjd}+H_jZ_{rd}+M_{rjd}+L_{rjd}+U_{jd}\right)$$
$$+\sum_p\sum_j\sum_i I_{pji}\left(D_{pji}+H_jZ_{pi}+M_{pji}+L_{pji}+U_{ji}+K_{pji}+Q_{pji}\right)$$
$$+\sum_i\sum_j\sum_d S_{ijd}\left(E_{ijd}+H_jZ_{id}+M_{ijd}+L_{ijd}+U_{jd}\right)$$
$$+\sum_p\sum_j\sum_m W_{pjm}Y_m\left(F_{pjm}+H_jZ_{pm}+M_{pjm}+L_{pjm}+U_{jm}+K_{pjm}\right)$$
$$+\sum_m\sum_j\sum_r X_{mjr}Y_m\left(G_{mjr}+H_jZ_{mr}+M_{mr}+L_{mjr}+U_{jr}+V_{mj}\right)$$
$$+\sum_m Y_mP_m \qquad (4.22)$$

Subject to:

$$\sum_p O_{pjd}+\sum_r T_{rjd}+\sum_i S_{ijd}=N_{jd}\quad \forall j\in J \text{ at each } d\in DD \qquad (4.23)$$

$$\sum_p R_{pjr}+\sum_m X_{mjr}-\sum_d T_{rjd}=0\quad \forall j\in J,\forall r\in RR \qquad (4.24)$$

$$\sum_p I_{pji}-\sum_d S_{ijd}=0\quad \forall j\in J,\forall i\in IM \qquad (4.25)$$

$$\sum_p W_{pjm}-\sum_r x_{mjr}=0\quad \forall j\in J,\forall m\in MC \qquad (4.26)$$

$$\sum_p\sum_j W_{pjm}\ge MV_m,\ \text{or otherwise}=0\quad \forall m\in MC \qquad (4.27)$$

$$\sum_p\sum_j R_{pjr}\ge MV_r,\ \text{or otherwise}=0\quad \forall r\in RR \qquad (4.28)$$

$$\sum_p\sum_j I_{pji}\ge MV_i,\ \text{or otherwise}=0\quad \forall i\in IM \qquad (4.29)$$

$$\sum_m Y_m\le MAXMC \qquad (4.30)$$

The objective function of this formulation minimizes the total logistics costs incurred in delivering vehicles to dealers from plants and ports of entry. It also includes an amortized fixed cost to evaluate the potential cost of adding multilevel rail mixing centers to the existing rail and intermodal network. Constraints (4.23) assure that the annual demand for each vehicle family at each dealer is satisfied.

Constraints (4.24), (4.25) and (4.26) act as conservation of flow or balance type constraints. Specifically, constraints (4.24) guarantee that the total vehicles of each vehicle family shipped from plants and from multilevel mixing centers to a destination multilevel rail yard equals the number of vehicles of each family shipped from that destination multilevel rail yard to dealers. Constraints (4.25) assure that the total vehicles of each vehicle family shipped from plants to a destination intermodal yard equal the number of vehicles of each family shipped from that destination intermodal yard to dealers. Finally, constraints (4.26) stipulate that the number of vehicles of each family shipped from a plant to a rail mixing center must equal the number of vehicles of each family shipped from that rail mixing center to destination multilevel rail yards. Note that it is actually not necessary to sum over plants and ports in constraints (4.23), (4.24), (4.25) and (4.26) because we defined a unique vehicle family for each family at each plant. However, had we not imposed this restriction (to keep manufacturing sourcing fixed), the slightly more general form shown in (4.23), (4.24), (4.25) and (4.26) would be necessary. Constraints (4.27) imposed a minimum annual vehicle volume threshold on any potential multilevel rail mixing center, while constraints (4.28) and (4.29) placed similar restrictions on multilevel and intermodal destination yards, respectively. Constraints (4.30) limited the total number of rail mixing centers which could open and operate. Constraints (4.30) were often employed in addition to constraints (4.27) in modeling runs because certain rail operational considerations suggested that the number of mixing centers should be limited, regardless of whether minimum thresholds could be met at more potential mixing centers than this limit allowed.

Model Results And Case Study Summary

As is often the case in an applied study, results were generated for a wide range of alternative scenarios. The major classes of results generated included:

(1) current network solutions - *i.e.*, immediate improvement opportunities based on the current network infrastructure and available equipment,

(2) potential future containerized intermodal opportunities, and

(3) potential future mixing center opportunities

Briefly, this study succeeded in identifying significant savings and operational improvement opportunities both in the near term and for the longer term. The vehicle manufacturer employed this study's short term oriented modeling results to direct where to make immediate changes from conventional rail to intermodal rail. Additionally, major savings opportunities were identified with the mixing center concept and this stimulated the manufacturer and its rail transport partners to develop and proceed with an implementation strategy for mixing centers.

Final Notes On Case Study

This case study illustrates another example of the type of strategic distribution and transportation planning that one commonly finds taking place independent of strategic manufacturing planning activities. This case represented an extremely complex problem from both a modeling and practical implementation perspective even with manufacturing sources held fixed. Further, from an organizational point of view, it would have added yet another level of complexity to attempt to integrate a manufacturing group into this planning project. This is not to say that this additional level of integration might not have helped further. However, it is a tacit recognition of the organizational and political realities that affect the scope and level of integration of planning activities in many firms.

4.1.4 Selected Other Strategic Distribution And Transportation Problems

In closing section 4.1, we will briefly consider two additional strategic distribution and transportation network planning problems. First we review a problem sometimes termed the combined warehouse location and routing problem or alternatively, the depot and direct store delivery (DSD) problem. Then, we examine a planning approach to the design and operation of motor carrier LTL networks. We will observe that these problems typically require both a hierarchical planning and solution approach.

The Warehouse Location And Routing Or Depot And DSD Problem

For purposes of this section, we define the warehouse location and routing problem as the problem of both locating DC's (or depots) to serve local markets, while at the same time determining the size of the local delivery fleet and the actual routes required to serve customer locations in markets. A potential subset of this problem would be the problem of determining the depots and routes to serve markets, while leaving the actual ownership and sizing of a fleet to a third party provider.[6]

This problem differs from the traditional warehouse location model (section 4.1.1) in several ways. The warehouse location model typically focuses on locating national or regional DC's that will serve customers in multiple regions with direct linehaul shipments. In contrast, the warehouse location and routing problem focuses more on locating local depots that will serve customers with local routes, each of which serves multiple customers on one outbound and return trip from and back to the depot. Additionally, the traditional warehouse or depot location and routing problem supports the development of "peddle" runs (perhaps) daily from local depots to a set of local customers on each route. As noted, this problem also often has a fleet sizing dimension to it in addition to the warehouse location and routing dimensions typically found. The traditional warehouse location model does not include the determination of the composition and size of a delivery fleet as one of its decision support outputs. It also does not address "peddle" run type planning problems.

As noted, we will consider the warehouse location and routing problem as representing the problem of determining the optimal location or locations of a local warehouse(s) while simultaneously determining the actual routes and/or size of the delivery fleet that will operate out of these warehouses. Thus, we have defined this

[6] Whether this would really represent a subset of the first problem can be argued. However, we point this out to clarify that there exist variations of this problem that do not necessarily involve owning a fleet.

problem in the spirit of Perl and Daskin (1985) who developed and implemented a decomposition algorithm to solve this problem. As these authors point out, the warehouse location and routing problem contains elements of several well-known problems including the traveling salesman problem and the multi-depot vehicle dispatch problem. This problem possesses a hierarchical planning dimension in the sense that while the facility location portion of the problem represents a strategic decision, the fleet-sizing and particularly the vehicle routing portion of this problem can also be characterized as more tactical planning in nature. Further, once a network is established, actual routes may change at an operational executional level. An important contribution of this problem is that it recognizes the cost and service interdependence between the more strategic decision of locating facilities and the more tactical decisions of planning routes and/or fleets to serve routes. For purposes of illustration, we will now review the formulation and decomposition approach that Perl and Daskin (1985) created to address this problem.

Perl and Daskin begin by defining that there exist a set of N customers in a market with known demand requirements. Additionally, they define that there exist M potential locations at which to site depots to serve the set of N customers. The objective of their model is to minimize the fixed and variable warehousing costs, the trunk (or stem) transportation costs and the variable vehicle delivery costs. Trunk or stem transportation costs are defined as the sum of the costs for the delivery vehicle to move from its base depot to the first customer served on its route and to return from the last customer on its route to the depot. The variable vehicle delivery costs include the transport costs incurred in a route between the first customer location on a route and the last customer location on a route. Perl and Daskin employ the following notation to formulate their warehouse location and routing problem.

Table 4.5 Notation For A Warehouse Location And Routing Problem

(i) Subscripts:

h, g = "point" index (customer or DC site)
$(1 \leq h \leq N+M), (1 \leq g \leq N+M)$
i = customer index $(1 \leq i \leq N)$
j = DC site index $(N+1 \leq j \leq N+M)$
k = route index $(1 \leq k \leq K)$
s = supply source index $(1 \leq s \leq S)$

(ii) Sets:

$p = \{p_h | 1 \leq h \leq N+M\}$ set of all points
$W = \{w_j | N+1 \leq j \leq N+M\}$ set of all potential DC sites
$L = \{ w_j | Z_j = 1\}$ set of opened DC sites

Table 4.5 Notation For A Warehouse Location And Routing Problem (continued)

(iii) Parameters:

d_{ij} = distance between points i and j
q_i = requirement of customer i
FC_j = fixed cost of establishing DC j
VC_j = variable cost per unit throughput at DC j
T_j = maximum throughput at DC j
CT_{sj} = unit cost of trucking from supply source s to DC j
C_k = capacity of vehicle (or route) k
D_k = maximum allowable length of route k
CM_k = cost per mile of delivery vehicle on route k

(iv) Variables:

$$X_{ghk} = \begin{cases} 1 \text{ if point } g \text{ precedes } h \text{ on route } k \\ 0 \text{ otherwise} \end{cases}$$

$$Y_{ij} = \begin{cases} 1 \text{ if customer } i \text{ is allocated to DC } j \\ 0 \text{ otherwise} \end{cases}$$

$$Z_j = \begin{cases} 1 \text{ if customer } i \text{ is allocated to DC } j \\ 0 \text{ otherwise} \end{cases}$$

f_{sj} = quantity shipped from supply source s to DC j

Given this notation, the authors formulate the following mixed integer programming model which they term the warehouse location and routing problem.

A Warehouse Location And Routing Problem

$$\text{Minimize: } C(X,Y,Z,f) = \sum_{j=N+1}^{N+M} FC_j Z_j + \sum_{s=1}^{S} \sum_{j=N+1}^{N+M} CT_{sj} f_{sj}$$

$$+ \sum_{J=N+1}^{N+M} VC_j \left(\sum_{i=1}^{N} q_i Y_{ij} \right) + \sum_{k=1}^{K} \sum_{g=1}^{N+M} \sum_{h=1}^{N+M} CM_k d_{gh} X_{ghk} \tag{4.31}$$

Subject to:

$$\sum_{k=1}^{K} \sum_{h=1}^{N+M} X_{ihk} = 1 \qquad i = 1, \ldots, N \tag{4.32}$$

$$\sum_{i=1}^{N} q_i \sum_{h=1}^{N+M} X_{ihk} \le C_k \qquad k = 1, \ldots, K \tag{4.33}$$

$$\sum_{g=1}^{N+M} \sum_{h=1}^{N+M} d_{gh} X_{ghk} \le D_k \qquad k = 1, \ldots, K;\ h = 1, \ldots, N+M \tag{4.34}$$

$$\sum_{g \in V} \sum_{h \in \bar{V}} \sum_{k=1}^{K} X_{ghk} \geq 1 \; \forall (V, \bar{V}) \tag{4.35}$$

Where V is a proper subset of P containing W

$$\sum_{g=1}^{N+M} X_{hgk} - \sum_{g=1}^{N+M} X_{ghk} = 0 \quad k = 1,\dots,K;\ h = 1,\dots,N+M \tag{4.36}$$

$$\sum_{j=N+1}^{N+M} \sum_{i=1}^{N} X_{ijk} \leq 1 \qquad k = 1,\dots,K \tag{4.37}$$

$$\sum_{s=1}^{S} f_{sj} - \sum_{i=1}^{N} q_i Y_{ij} = 0 \qquad j = N+1,\dots,N+M \tag{4.38}$$

$$\sum_{s=1}^{S} f_{sj} - T_j Z_j \leq 0 \qquad j = N+1,\dots,N+M \tag{4.39}$$

$$\sum_{h=1}^{N+M} X_{ihk} + \sum_{h=1}^{N+M} X_{jhk} - Y_{ij} \leq 1 \qquad \begin{array}{l} i = 1,\dots,N \\ j = N+1,\dots,N+M \\ k = 1,\dots,K \end{array} \tag{4.40}$$

$$X_{ghk} = 0,1 \quad g = 1,\dots,N+M \qquad \begin{array}{l} h = 1,\dots,N+M \\ k = 1,\dots,K \end{array} \tag{4.41}$$

$$Y_{ij} = 0,1 \qquad i = 1\dots,N \qquad j = N+1,\dots,N+M$$

$$Z_j = 0,1 \qquad j = N+1,\dots,N+M$$

$$f_{sj} \geq 0 \qquad s = 1,\dots,S \qquad j = N+1,\dots,N+M$$

Briefly, this model works as follows. The objective function (4.31) minimizes the fixed and variable warehousing costs, the trunk (or stem) transportation costs and the variable vehicle delivery costs. Constraints (4.32) assure that all customers are served and that no customer appears on more than one route. Perl and Daskin also point out that this constraint implies that a customer's demand level cannot exceed the capacity of the vehicle serving the route. If a customer's demand does exceed the delivery vehicle's capacity, then they assume that this is determined in advance and that direct truckload service from the depot (DC) to the particular customer is established to circumvent this capacity problem.

Equations (4.33) define the capacity constraints of each of the k routes or delivery vehicles, while constraints (4.34) specify the maximum allowable length (or distance) of a route. The authors employ constraints (4.34) to assure that any particular route does not require a driver to be on the road beyond some defined time limit. Thus, the total route length serves as a surrogate for hours required to complete a route in this formulation. Constraints (4.35) associate each delivery route with a particular DC, while constraints (4.36) essentially act as a balance equation stipulating that any geographic point that a delivery vehicle enters must be exited by that same vehicle (or route). Equations (4.37) assure that each delivery route operates out of only one DC. Equations (4.38) and (4.39) represent fairly traditional constraints which stipulate that the flows into a DC must equal the flows out of a DC, and that flows through a DC cannot exceed a DC's capacity. Constraints (4.40) in conjunction with the constraints imposed by equations (4.32) and (4.35) - (4.37) specify that a customer can only be served by a DC if there exists a route which originates from that particular DC and which goes through the particular customer.[7] Finally, constraints (4.41) define the 0-1 binary variables and non-negativity flows required for this problem.

As previously noted, this problem has an implicit hierarchical dimension to it in that while the location of warehouses typically represents a strategic or perhaps tactical decision, the development of routes to serve customers typically represents an operational or tactical decision. The warehouse location and routing problem thus spans a hierarchy of planning horizons in an attempt to address the interdependence between the decision of where to locate depots to serve a market and the decision of how to develop routes and/or use a fleet of delivery vehicles to serve local customers at the operational executional level.

Solution approach represents the other hierarchical aspect of the warehouse location and routing problem. This problem as originally defined by Perl and Daskin in 1985 represented (and still represents) a difficult mixed integer problem to solve. Thus, the authors implemented a clever hierarchical or decomposition approach in order to use this model for applied problems. The reader interested in the details of the decomposition approach employed is referred to Perl and Daskin (1985). Briefly, however, the authors make several simplifying assumptions regarding the original problem, and then decompose this modified problem into three subproblems:

1. A multi-depot vehicle dispatch problem,
2. a warehouse location-allocation problem, and
3. a multi-depot routing-allocation problem

Sub-problem (1) assumes that all *potential* depot sites described in a problem are operated and then develops a preliminary set of routes to minimize total delivery costs. From the original set of potential depot locations, sub-problem (2) then determines the actual warehouse locations that will operate. This sub-problem also

[7] For a more detailed explanation of how these constraints collectively impose this requirement, see Perl and Daskin (1985).

assigns the routes designed in sub-problem (1) to the depots opened in sub-problem (2) - *i.e.*, the route allocation process. Finally, sub-problem (3) simultaneously reassigns the customers to the depots and determines a solution to the multi-depot routing problem for the warehouses opened by sub-problem (2).

This concludes our review of the warehouse location and routing problem. It has served to illustrate yet another example of the hierarchical planning and solution characteristics commonly found in distribution and transportation planning. We will now consider another network infrastructure and routing problem, namely, the motor carrier LTL network design and operation problem.

A LTL Motor Carrier Network Problem

LTL network design and scheduling problems have received a great deal of attention both in research and applied settings (see e.g., Crainic and Roy [1988], Daganzo [1987], and Powell and Sheffi [1986, 1989]). To facilitate a discussion of these problems, Figure 4.6 illustrates a small LTL network. Briefly, a LTL network consists of a series of end-of-line terminals and breakbulk terminals. End-of-line terminals function as shipment origin and destination locations, while breakbulk terminals function as transshipment and consolidation locations. Individual LTL shipments generally range from a minimum of several hundred pounds up to 10,000 pounds or more. LTL carriers consolidate multiple shippers' smaller shipments and transport them across networks in consolidated truckload quantities (*i.e.*, consolidated shipments typically weighing up to 40,000 pounds or more). LTL carriers primarily move shipments from an origin to a destination by a series of one or more linehauls that generally flow through the breakbulk terminals. At each breakbulk terminal a sort and segregate operation occurs and individual shipments then move on to the next terminal on a linehaul move. For example, a shipment originating from end-of-line terminal 1D and destined for end-of-line terminal 2D in Figure 4.6 would flow through both breakbulk terminals BBT1 and BBT2 on the way to end-of-line terminal 2D. In addition, LTL carriers' networks typically have a limited number of point-to-point direct moves over certain high volume lanes. The dotted line between end-of-line terminal 1A and breakbulk terminal 2 in Figure 4.6 illustrates a direct service link in an LTL network. In deciding upon whether to establish a direct move between two terminals, a critical factor that a LTL carrier evaluates is the frequency of shipments that its normal volume will allow between two terminals. Usually, a carrier will not consider operating a direct link between two terminals unless it can consistently generate a minimum of three full truckloads or more per week. The routing of empty trucks to balance the flow of trailers over lanes represents another critical element in the successful operation of an LTL network. Empty moves do not generate revenue and thus carriers seek to minimize such moves while still positioning (*i.e.*, balancing) trailers to meet shippers demands. Finally, many LTL carriers also offer some direct truckload service, and these carriers must determine how to effectively integrate their truckload service with their LTL service. All of the preceding LTL network planning elements collectively combine to form what researchers have termed the "load planning" problem (see e.g., Powell and Sheffi, 1989).

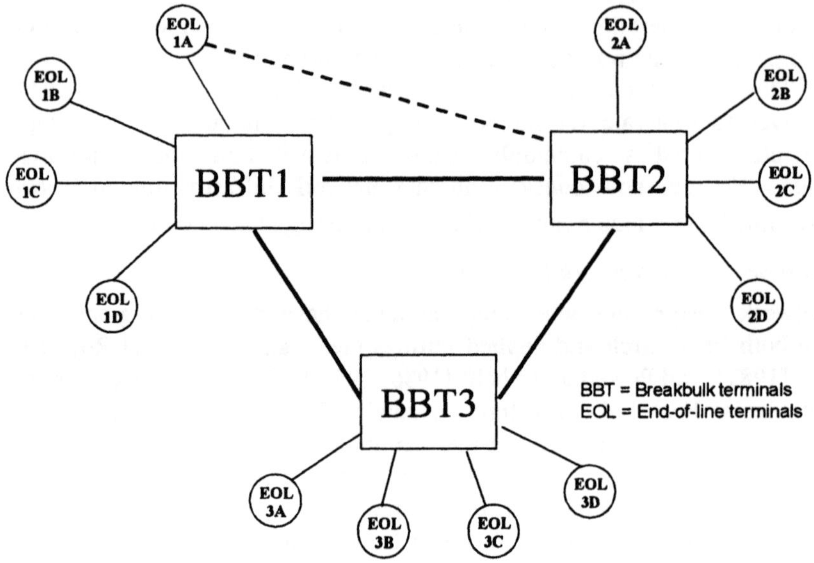

Figure 4.6 Illustrative LTL Network

To summarize, the LTL load planning problem consists of:

1. determining the origin-destination terminal pairs between which to offer direct point-to-point service
2. routing freight over the defined LTL network using available links between breakbulk terminals and their respective satellite end-of-line terminals, as well as available point-to-point direct links,
3. routing empty trailers over the network as appropriate to ensure that customers demand can be satisfied, and
4. integrating and routing truckload freight along with the LTL freight (for those LTL carriers who handle truckload volume).

The objective of the load planning problem is to minimize the total cost of the transportation and handling operations associated with operating the network including all flows (both loaded and empty) and all unloading, loading, sorting and segregating activities, subject to the customer service requirements and constraints of the network. Before describing a well known approach by Powell and Sheffi (1989) to this problem, we first consider the hierarchical characteristics of the load planning and LTL network design and scheduling problem. (Note that the term LTL network design problem generally refers to only the problem defined by the first two elements of the four element load planning problem just defined. The load planning problem is typically referred to as the full set of four elements just defined.)

Hierarchical Characteristics Of The LTL Network Design And Scheduling Problem

The LTL network design and scheduling problem shares a number of the hierarchical dimensions previously observed in hierarchical production and distribution planning problems. At the strategic planning level, an LTL carrier must determine the long-term infrastructure that it will employ to support its tactical plans and actual operations. In particular, a carrier must make such decisions as what geographic regions its operations will serve, where it will locate its breakbulk and end-of-line terminals, and the size of the fleet that it will operate. One can argue that this latter decision (*i.e.*, size of fleet) represents somewhat of a more tactical than strategic decision. Regardless, these major strategic infrastructure decisions create both opportunities and constraints at the tactical and operational planning and scheduling levels for an LTL carrier, in an analogous manner to the way that strategic infrastructure decisions generate opportunities and constraints at lower planning levels for manufacturing companies.

At the tactical planning level, a LTL carrier must evaluate how to most effectively utilize its infrastructure and also make minor adjustments to its infrastructure. For example, at this level, a carrier typically would determine over which lanes it should offer point-to-point direct service between a breakbulk terminal and an end-of-line terminal (which is not a satellite of a particular breakbulk terminal). A carrier will also typically review its tractor and trailer fleet size at the tactical planning level to determine if it needs to adjust its existing fleet. Additionally, a carrier also generally develops basic guidelines for maintaining a balanced deployment of trailers across its network at this planning level.

At the operation planning and scheduling level, LTL carriers again perform activities very analogous to those conducted by manufacturers operating a hierarchical production planning system. Carriers schedule the daily movements of their fleet of tractors and trailers hauling shippers' freight over the network. In some cases, sophisticated optimization-based scheduling systems provide decision support to guide freight scheduling activities just as optimization models schedule production lines for manufacturers. Carriers must also address short run capacity issues that may suddenly develop at the operational level. Similarly, they must determine plans on a weekly and daily basis as to how "essentially" to deploy inventory over the network. Deploying inventory for LTL carriers can be thought of as deciding where to position empty trailers in order to serve short run service orders and/or projected shipper demands.

Similar to manufacturers, LTL carriers must also incorporate feedback loops into their hierarchical planning systems. This ensures that bottlenecks, constraints, costs and other operational indicators flow back up to the higher planning levels. For example, a particular breakbulk to breakbulk linehaul or point-to-point linehaul from an end-of-line terminal to a breakbulk terminal may not be generating the expected volume on a lane. This information must flow back to the tactical planning level from the operational level so that appropriate schedule and/or route adjustments can be implemented to reduce costs and/or improve service. In such a case, it may become necessary to either increase or reduce the number of lanes serviced by direct point-to-point linehauls in response to actual volumes at the operational level. Similarly a carrier may have to either increase or decrease the number of tractors and/or trailers in its fleet in response to actual demands at the operational level. These all represent the types of decisions that must be addressed from a tactical planning perspective, and feedback loops from the operational planning level to the tactical level facilitate agile and effective decision-making in such situations. In summary, LTL carriers face many very similar issues to those of manufacturers in designing and executing a true hierarchical planning system to support their operations.

Solution Approach For The Network Design And Load Planning Problem

The full network design/load planning problem represents a very complex mathematical problem. Sheffi and Powell (1989) developed and implemented a now well known approach to this problem for a carrier known as PIE Nationwide in the late 1980's. The reader interested in details about the innovative formulation and algorithmic approach of these authors is referred to the aforementioned citation. We will now focus at a summary level upon the authors' approach.

In describing their decision to utilize a decomposition approach, Sheffi and Powell note that:

> The natural structure of the problem, as usual in transportation, lent itself to a hierarchical solution approach that simplified the algorithms, as well as significantly accelerated their speed.... The problem [can] be decomposed hierarchically into a network design problem, governing which links are added or dropped from the network, and a series of subproblems determining the routing of LTL shipments, the routing of TL [truckload] shipments and the routing of empty trailers to balance the network.

The authors also noted that their decomposition approach "was useful in separating the more accurate cost calculations from the less accurate ones". This concept is related to an earlier theme that we noted about hierarchical production planning; namely, that by separating the production planning process into strategic, tactical and operational planning phases, a firm can delay making decisions until they (the decisions) are required to facilitate the production process. Recall that by delaying decisions until absolutely required, we observed that decisions could be made more accurately and/or with less risk.[8]

Sheffi and Powell's hierarchical approach to solving the four decisions of the load planning problem consisted of decomposing the problem into the following components:

1. a network design problem (*i.e.*, the problem of determining the network),
2. the routing subproblem,
3. the truckload freight routing problem, and
4. the empty balancing subproblem

---

[8] For example, delaying the decision about what end item to produce until very shortly before actual production represented a classic example of delaying (or decomposing) the production decision-making process.

The network design problem determines all the nodes or origin-destination pairs between which a LTL carrier will offer direct service (*i.e.*, the nodes between which the carrier will operate its fleet). The solution to this problem defines what regions the carrier can serve. Whenever, a carrier updates its network design problem, a possible outcome is that the carrier may add or drop links between terminals on its existing network and/or add or drop terminals from its network. The network design decision must evaluate all of the transportation and handling costs associated with operating a network, as well as service considerations.

A solution to the network design problem facilitates solving the three sub-problems defined by Sheffi and Powell. The routing subproblem determines the optimal assignment (*i.e.*, routing) of all LTL flows over the defined network including the volume of freight that will flow through each breakbulk terminal. As its name connotes, the truckload freight routing problem assigns the truckload moves that will flow over the LTL carrier's network. The reason that the truckload routing problem must be considered in this hierarchical problem is that truckload moves do affect the deployment of the carrier's trailers over its network; and therefore, the impact of truckload moves must be considered in the empty balancing subproblem. The empty balancing subproblem routes the LTL carrier's empties over its network, as necessary, in order to correct any surpluses or deficits of actual or projected available trailers at all nodes across the network. This purpose of this subproblem is to assure that the LTL carrier positions its empty trailers appropriately across its network in order to serve shipper demands as efficiently and effectively as possible. The routing subproblem and the empty balancing subproblem in particular are dependent upon the network design subproblem. That is to say, solutions to these two respective subproblems are essentially a function of the network design problem. Therefore, whenever the network design problem is re-optimized, these two problems must always be optimized.

In summary, the LTL load planning and network design problem represents yet another distribution and transportation problem that has a strong hierarchical dimension to it. Similar to previous problems studied in this and preceding chapters, in its totality, this problem addresses a myriad of issues ranging from long run infrastructure decisions to short run scheduling decisions.

4.2 Short Run Transportation Scheduling

The topic of short run, operational transportation planning and scheduling has received extraordinary attention in both academic and research settings as well as applied settings. Today there exist numerous commercial software packages and systems that provide optimization based decision support in this area. Additionally, researchers and industry practitioners have also implemented untold numbers of customized scheduling systems. The forms of transportation planning and scheduling that abound range from the scheduling of shipments by manufacturers using third party commercial transport providers, to the scheduling of large scale national and international fleets, to the scheduling of transportation fleets serving a particular market with "peddle runs" from one or more local depots, and so on.

In this section, we will explore further how transportation scheduling fits into the hierarchical planning process of manufacturers and distributors operating global and domestic supply chains. Following the theme of this book, we will focus in particular on certain hierarchical dimensions of the short run transportation planning and scheduling process. A review of this topic would be remiss, however, without first recognizing one of the central underlying problems in transportation scheduling; namely, the traveling salesman problem and the closely related vehicle scheduling problem.

The *traveling salesman* problem consists of determining the minimum cost single route for one vehicle (and salesman) to serve a fixed set of customers. The route must visit each customer exactly once and conclude at its original starting point. Table 4.6 and model (4.42 - 4.46) illustrate a typical integer programming formulation for the traveling salesman problem.

Table 4.6 Notation For The Traveling Salesman Problem

L	is the number of customer locations to be served by the salesman/vehicle.
i and j	are indices to represent customer locations. Note that the starting and ending point of the route, e.g., the depot, is location 0. i or j = 1,. . . , L represents the actual customer locations.
c_{ij}	is the cost to travel between customer i and customer j
$x_{ij} =$	$\begin{cases} 1, \text{ if the salesman/vehicle travels directly from customer } i \text{ to customer } j \\ 0, \text{ otherwise} \end{cases}$

The Traveling Salesman Problem

Minimize:
$$\sum_{i=0}^{L}\sum_{j=0}^{L} c_{ij}x_{ij} \tag{4.42}$$

Subject to:
$$\sum_{i=0}^{L} x_{ij} = 1 \quad \text{for all } i \tag{4.43}$$

$$\sum_{j=0}^{L} x_{ij} = 1 \quad \text{for all } j \tag{4.44}$$

$$\sum_{i\in s}\sum_{j\in \bar{S}} x_{ij} \geq 1 \quad \text{for each proper non-empty subset S of } \{0,1,\ldots,L\} \tag{4.45}$$

$$x_{ij} = 0 \text{ or } 1 \quad \text{for all } i \text{ and } j \tag{4.46}$$

The cost of traveling the route in the objective function (4.42) can be defined using alternative measures such as distance, time or actual costs. Equations (4.43) and (4.44) constrain the optimal route to visiting each customer one time on the tour, while constraints (4.45) assure that a fully connected optimal tour results.

The *vehicle routing problem* essentially represents an extension of the traveling salesman problem for the case where a vehicle or salesman traveling a single route cannot cover all of the fixed customer locations that require a visit. In this case, multiple routes originating from one origin (e.g., a depot) are required to visit the set of the customer locations to be served. Similar to the traveling salesman problem, the objective of the vehicle routing problem consists of determining the cost minimizing set of routes to serve a fixed set of customers from a single depot or origin.

Figure 4.7 provides a simple illustration of the type of solution that results from a vehicle routing problem. It shows a solution for a hypothetical example in which the individual delivery vehicles have a capacity of 500 units and each customer location has a forecasted or assumed average daily demand. These represent just a few of the factors that can enter into the determination of an optimal set of routes to serve a defined set of customers from a single depot. Time constraints on the duration of the total route, time windows for delivering to individual customers, vehicle constraints (e.g., maximum size vehicle that certain customer locations can receive), etc... represent several more of the numerous constraints that practical vehicle scheduling algorithms may have to consider.[9] Extensions to the single depot vehicle routing problem naturally include the multi-depot vehicle routing problem. The reader interested in additional background on traveling salesman and vehicle routing problems, and vehicle scheduling in general, is referred to the following references on this topic as a starting point (Clarke and Wright, 1964; Dantzig and Ramser, 1959; Desrochers, Lenstra and Savelsbergh, 1990; Fisher and Jaikumar, 1981; Golden and Assad, 1988; Golden, Laporte and Taillard, 1997; Laporte, 1992; Powell and Mayoras, 1997; Powell, Sheffi, et. al. 1988; Ronen, 1988; Savelsbergh and Goetschalckx, 1995; and Solomon, 1986 and 1987).

With this brief review of the traveling salesman and vehicle routing problem completed, we return now to the role of transportation scheduling in a hierarchical planning context. In this discussion, we focus on truck routing and scheduling. Ronen (1988) in a survey article on truck routing and scheduling put this activity into perspective noting that:

> Truck routing and scheduling is not done in a vacuum. Strategic and tactical decisions set the stage for truck routing, and truck routing and scheduling imposes requirements on driver scheduling, and truck loading/unloading schedules.

[9] See Ronen (1988) for a quite comprehensive classification and review of the types of constraints often required in vehicle routing and scheduling problems.

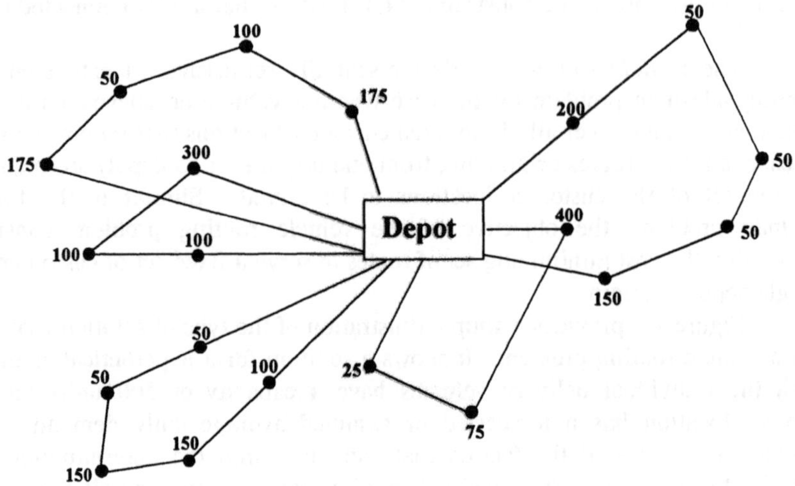

- The numbers at each node represent the average daily demand at that customer location.
- This example assumes that each delivery vehicle has a capacity of 500 units.

Figure 4.7 Illustration Of A Vehicle Routing Solution

Thus, as Ronen points out, we must view truck routing, and more generally, short run transportation planning and scheduling in its proper hierarchical perspective. The strategic and tactical manufacturing and distribution planning decisions previously implemented by a firm significantly affect short run transportation scheduling and operations. The network infrastructure (e.g., plants, distribution centers and depots) and any owned transportation assets (e.g., the composition of a private fleet) both provide capabilities and impose constraints on transportation operations. For example, the location of depots and the number of vehicles in a firm's private delivery fleet will affect actual schedules created to serve a local market. From a hierarchical planning perspective, it is important to consider the impact of higher level decisions on transportation operations and scheduling. One response to the recognition of the need for this has been the attempts by researchers to develop models to simultaneously evaluate network infrastructure and transportation planning/scheduling decisions. The combined warehouse location and routing problem of Daskin and Perl (1985) discussed in section 4.1 represents one example of such an attempt. The work of Golden, Assad and Dahl (1984) to explore the relationship between fleet size and composition, and vehicle routing schedules and costs represents another well known example of such an effort.

An important point to note is that in integrating tactical and/or strategic planning with operational planning in one comprehensive modeling approach, it is generally necessary to assume "averages" for the operational transportation planning component of the model. Averages generally take the form of a forecasted average customer delivery requirement or demand by whatever appropriate time bucket (e.g., daily, weekly, monthly or annually). For example, in the vehicle routing solution illustrated in Figure 4.7, an average daily demand or delivery quantity at each customer location is assumed. A key issue in real world operational scheduling is that one usually finds significant variability in the daily demands for transportation services as customer orders can vary widely by day, week and so on. Further, it is very difficult to accurately capture this "variability" factor in large scale decision support models spanning multiple planning horizons (e.g., tactical and operational). Thus, addressing this real world consideration represents a major challenge in integrating short run operational factors into planning models.

One "proactive" approach to this problem is to simulate operational transportation scheduling within the context of an overall hierarchical modeling approach. For example, consider the combined warehouse location and routing model reviewed in section 4.1. The output of this optimization model would consist of a recommended set of depots and routes to serve one or more markets. Assume that a firm plans to use this model's recommended solution as a blueprint to construct a network of depots. This would represent a major investment by a firm and it may wish to explore in more detail how efficiently it could utilize the proposed network infrastructure at the operational level (e.g., on a day-to-day or week-to-week basis). To do this, a firm could construct an operations scheduling simulation model to complement its warehouse location and routing model. Under this planning approach, the following steps would occur:

Step 1: The firm runs its warehouse location and routing model to determine a recommended set of depots and routes to serve a market.

Step 2: Based on history and/or forecasts, the firm develops a data set of projected daily demands by customer location for an appropriate, extended period of time (e.g., 250 days).

Step 3: The firm simulates the projected daily activity on its proposed depot network capturing such key results as "simulated" actual costs vs. projected (based on averges), variances in daily capacity utilization rates, the simulated frequency of the need to deviate from planned routing schedules, and so on.

The simulation analysis described in step 3 could be applied to several alternative depot infrastructures developed using sensitivity analysis on the results of the combined warehouse location and routing model. These simulation results would serve to provide an "operational" perspective or "reality check" on the initial network infrastructure plans yielded by the higher planning level optimization model. It would not be unusual to find that this additional analysis could lead to a different final judgement than a firm might arrive at just based on tactical and/or strategic modeling analyses.

Simulation therefore represents an important option to consider in developing a hierarchical approach for both *planning* the construction of, and then *assessing,* on an ongoing basis, the infrastructure and transportation links and schedules of a supply chain network. In particular, simulation can provide insights *in advance* as to how potential tactical (and even strategic) infrastructure decisions may impact real world transportation operations. Thus, operational level simulations facilitate the critical feedback necessary from the operations to the tactical and strategic planning levels.

A second, more traditional form of feedback from the short run transportation scheduling level takes the form of monitoring operations and reporting results back to the tactical planning levels on a regular basis. If this feedback of operating results indicates the need for a modification of tactical plans, then the tactical planning processes in place must address the situation. We previously observed an example of this feedback mechanism in the multi-mode, multi-echelon problem reviewed in section 4.1.2. In particular, Table 4.3 displayed illustrative operating results for the transportation lanes of this network. Recall that this example showed very poor operating conditions on one lane and this feedback to the tactical planning level would have dictated modifications to the tactical plan. Under ideal planning and operating conditions, a firm will have in place both "proactive" mechanisms such as simulation, as well as ongoing - more reactive- mechanisms for providing feedback between the transportation operations level and the network tactical planning level.

The Hierarchical Algorithmic Characteristics Of Short Run Transportation Scheduling

A rich and ever-growing literature on transportation and vehicle scheduling problems exists, and the boundaries and scale of vehicle scheduling problems that have surfaced in practical applications continues to expand. In this section we will briefly touch upon the algorithmic aspects of these problems, and the reader interested in exploring this subject further is referred to the previous citations in this chapter.

A very common approach to solving vehicle routing and scheduling problems is the hierarchical decomposition approach. In an interesting article written at a time when advances in vehicle routing and scheduling were occurring quite rapidly, Dempster et. al. (1983) noted the following.

> Many organizations operate vehicle fleets to deliver their products to customers. Frequently, for administrative convenience and other reasons, the fleet is scheduled using a fixed route system. In a fixed route system, particular customers are assigned to a vehicle and this assignment is revised infrequently (e.g., every 6-12 months). Daily customer orders are random (and) on any given day, a delivery schedule must be developed for the customers that have ordered for each vehicle.... This problem suggests an obvious hierarchical system in which one model is used to assign customers to vehicles to form the fixed routes, and another model is used to solve the daily scheduling problem ("cluster first, route second").

The suggested algorithmic approach posed by Dempster et. al., and many other researchers and practitioners over the years recognized several factors. First of all, routing and vehicle scheduling problems are inherently very difficult mathematical problems to solve. Therefore, decomposing any particular routing and scheduling problem into linked sub-problems often facilitates the development of more tractable formulations that one can consistently solve in acceptable solution times for operational use. Additionally though, this decomposition approach also often fits better from a practical, real world perspective. Specifically, daily delivery operations in many industries are very time sensitive (e.g., they often have tight delivery windows), and are subject to daily operational problems (e.g., congested traffic or congested customer receiving and unloading facilities). To successfully meet regular delivery schedules while circumnavigating daily impediments to delivery operations, firms often find that they must have route drivers who are familiar with a particular route or routes. This familiarity allows a driver to run a route efficiently making daily adjustments as necessary. Additionally, a driver's familiarity with customers and their product demands also can lead to more accuracy in the overall delivery process. In order to assure that drivers are familiar with their routes and customers, it is often necessary to establish fixed routes that have an assigned set of customers. Daily demands of customers will then drive the particular schedule on a route from day-to-day, however, the basic route remains fairly set over a significant period of time. Dillmann et. al. (1996) described the importance of these real world factors in their review of vehicle routing and scheduling approaches for serving magazine and newspaper wholesalers in Germany. These authors noted that in deciding between "fixed" and "variable" routes in this industry, practical operational aspects, in addition to algorithmic factors suggested the use of fixed rather than flexible or variable routes. In summary, mathematical tractability has often led practitioners and researchers to adopt a hierarchical decomposition approach to solving vehicle routing and scheduling problems. However, real world considerations often lead to a similar approach on non-mathematical grounds.

4.3 Tactical And Annual Mode Choice Planning

An important element of tactical (and annual) distribution and transportation planning is transportation mode choice selection. We have already examined mode choice planning in the context of integrated transportation and distribution planning in section 4.1. In this section, we reconsider mode choice selection problems from a slightly different perspective. Specifically, we present a methodology for making a modal choice decision on a transportation lane where one or more products are shipped between an origin and a destination. This method addresses the question of what transport mode should be used as the normal or standard mode for shipments between an origin-destination pair. Thus, this technique fits into the tactical planning category. This methodology has relevance to hierarchical planning for several reasons including:

- it employs an integrated approach where the impact on total supply chain costs are evaluated in the "freight" modal decision,

- it requires the development of "product group" aggregations in order to use it for multiple products, and
- it *can* employ a "hierarchical" management science planning technique known as the Analytic Hierarchy Process (Saaty, 1996) for decision-making support.

The analytic hierarchy process (AHP) has many potential applications in hierarchical operations and supply chain management. An exploration of this modal choice methodology will thus also serve to facilitate an illustration of the AHP process.

To demonstrate this section's modal decision-making technique, we pose a hypothetical case study of a firm deciding upon the proper transport mode to replenish a distribution center (DC) from a manufacturing plant. This methodology includes both a stand-alone transport mode choice costing technique, as well as a combined qualitative and quantitative (i.e., costing) approach. The AHP provides the mechanism for incorporating qualitative factors into the decision-making process.

In order to provide an introduction to the AHP for those readers interested in this technique, the following case study describes the development of an AHP model in some detail. For the reader less interested in the AHP or less interested in incorporating qualitative factors into the mode choice decision, a review of just the costing component sub-section of the following case will introduce the key concepts of this section (see Tables 4.7 and 4.8 in particular).

4.3.1 Case Study Of A Tactical Transport Mode Choice Decision[10]

Introduction

As manufacturers and distributors intensify their focus on their system-wide inventories, the trade-offs between utilizing air and ocean inventory pipelines (or motor carrier and rail transport pipelines), and the impact of mode and carrier selection on the entire distribution network continues to receive increasing scrutiny. Important cost considerations to evaluate in mode choice decisions include the trade-offs between inventory carrying costs, inventory investment costs and transport costs. Additionally, however, there are important qualitative factors that one must also consider in making this decision. For example, does a particular transport mode or individual carrier provide particularly good or bad customer service? And for that matter, what does customer service really mean to an individual shipper or transport carrier customer (i.e., how does an individual shipper define customer service)? Such factors should be evaluated systematically in the mode and carrier selection process. Without a suitable and proven evaluation framework, attempting to weigh quantitative and qualitative factors can become a rather clouded process subject to considerable speculation and inconsistency. This hypothetical case study demonstrates how one can use the

[10] This section is adapted from Liberatore and Miller, 1995.

analytic hierarchy process to incorporate both costs and a set of well-defined qualitative factors into a unified, quantitative evaluation system. Additionally, the cost methodology presented here can act as a stand-alone decision support tool, if there is no need to "quantitatively" incorporate qualitative factors into the decision-making process.

Problem Background

For illustrative purposes, assume that a firm wishes to evaluate the following set of cost and qualitative factors in making a modal and carrier choice decision. Figure 4.8 displays five major cost factors which will differ depending upon the mode and carrier that a firm selects to provide freight transport services. The cost factors shown are common to many shipping scenarios, and in particular, are extremely relevant to the case where a firm is shipping either finished goods or materials from one company location to a second company location.

1. Freight costs
2. The inventory carrying costs of inventory in the pipeline (i.e., in-transit inventory)
3. The inventory carrying costs of cycle stock at the receiving location (e.g., a logistics center)
4. The inventory carrying costs of the required safety stock at the receiving location
5. The investment cost required to produce the inventory to fill the pipeline

Figure 4.8 Cost Factors To Evaluate

The first four cost factors are annual recurring costs which will continue as long as the transportation inventory pipeline exists, while the fifth cost represents a one time cost required to initiate the pipeline. An example later in this case will briefly illustrate a quantitative methodology for evaluating these cost factors, and the reader is referred to Miller (1991A, 1991B and 1993) for a more detailed explanation.

On the qualitative side of the mode and carrier selection process, a large number of factors exist that a shipper might consider. Again for illustrative purposes, Figure 4.9 presents a set of qualitative criteria that the author has frequently observed shippers using in actual practice.

1. Perceived quality of customer services
2. Shipment tracking and tracing capabilities
3. Billing/invoicing accuracy
4. Electronic Data Interchange (EDI) capabilities
5. Potential to develop mutually beneficial long-term partnership
6. Cargo capacity limitations (if any)
7. Ability to provide service that does not damage goods while in transit
8. Customs clearance capabilities (for international shipments)
9. Impact on the shipper's negotiating position/leverage on other shipping activities

Figure 4.9 List Of Potential Qualitative Evaluation Criteria

Now consider the transport decision-maker attempting to determine the best mode and carrier to select, on the basis of both cost factors (Figure 4.8) and qualitative factors (Figure 4.9). In certain cases, the cost factors may so decidedly favor a particular mode and carrier that the decision becomes very straightforward. In many cases, however, one alternative will not readily stand out above all others, and the firm or decision-maker must somehow rationalize all important criteria in making a transport selection. The AHP represents a good multi-criteria decision-making methodology to employ for selecting the best transport mode and carrier.[11]

The Analytic Hierarchy Process (AHP)

The AHP developed by Saaty (1996) has effectively structured many types of complex, multicriteria business problems. It offers a flexible evaluation framework and an intuitive, easy-to-use approach for collecting subjective judgments. Equally as important, the AHP provides a measure of the level of consistency of the judgments collected. This consistency check may indicate that revision of certain judgments may be needed, and in this way it can help to improve the reliability of the evaluation process.

The AHP enables the decision maker(s) to rank alternative courses of action based on judgements concerning the importance of the criteria and the extent to which they are met by each alternative. These judgements are expressed in terms of pairwise comparisons of items on a given level of the hierarchy with respect to their impact on the next higher level. These pairwise comparisons are quantified using the standard one-to-nine AHP measurement scale (Figure 4.10).

Degree of Preference	Definition
1	Equally preferred
3	Moderately preferred
5	Strongly preferred
7	Very strongly preferred
9	Extremely preferred

Notes:

1. The intermediate judgments of 2, 4, 6 and 8 can be used as necessary.
2. The reciprocal judgements are interpreted as follows: if, for example, A is judged to be moderately preferred to B, the pairwise comparison between A and B is 3, implying the pairwise comparison between B and A is 1/3.

Figure 4.10 AHP Measurement Scale

In general, an AHP evaluation hierarchy contains a minimum of three levels. Two approaches for structuring the hierarchy are as follows:

[11] Other possible multi-criteria methods to consider include multi-attribute utility theory (see e.g., Keefer, 1978), scoring models (see e.g., Dean and Nishry, 1965), and goal programming (see e.g., Keown, Taylor and Duncan, 1979).

APPROACH 1: Standard AHP
(1) The goal,
(2) The criteria, and
(3) The alternatives

APPROACH 2: AHP with Rating Scales
(1) The goal,
(2) The criteria,
(3) The rating scale, and
(4) The alternatives

For the standard AHP approach, the process begins by pairwise comparing each alternative with respect to a specific evaluation criterion. For example, an ocean carrier would be compared to an air carrier concerning their respective response times to a shipper's request for service. A question might be stated as follows: with respect to response time, how much more important or better are carrier 1's capabilities as compared to those of carrier 2? If the response is moderately preferred, a 3 is the appropriate judgement. If there are four carriers being evaluated, then six pairwise comparisons are needed to round out the set of required response time judgments (i.e., 1 - 2, 1 - 3, 1 - 4, 2 - 3, 2 - 4, 3 - 4).

The pairwise comparison information is used to determine the weight or priority of each carrier. The same process of collecting and processing pairwise comparison data would then be repeated for each of the other evaluation criteria. This completes the first stage of the analysis.

In the second stage, the evaluation criteria themselves would be pairwise compared according to their perceived importance in selecting the best carrier. A typical question might be: how much more important is response time than tracking and tracing services in selecting the best carrier? This group of pairwise comparisons results in a set of weights for each evaluation criterion. The results of both stages would be combined using a weighted averaging technique to provide the final priorities or weights for each of the carriers.

The rating scale approach is similar to the standard AHP method except that a rating scale is introduced for each of the criteria. For example, specific performance ratings for tracking and tracing services would be defined (these are in fact used later and shown in Figure 4.11), and would be pairwise compared to determine their relative importance. For example, a question might be: how much better is an "outstanding" rating for response time as compared to an "above average" rating? At this stage all of the necessary pairwise comparisons would be collected and processed for the performance ratings. Next, the criteria would be pairwise compared as in the standard approach, and the weighted averaging technique would provide final weights for each rating level under each criterion. To complete the analysis for a specific carrier, an evaluator would assign one of the rating levels for each of the criteria. Summing up all of the weights associated with the selected rating categories provides a total score for each carrier. In this way, the AHP is used to generate the weights for what is essentially a scoring model.

Sub-Criterion: RESPTIM	
Rating	**Definition**
OUTSTAND	Immediate response or within 2 hours of the shipper's request.
ABVAVE	Same day response, if request received before noon.
AVERAGE	Response within 24 to 48 hours.
BELOWAVE	Response after more than 48 hours.

Sub-Criterion: RELIABL	
Rating	**Definition**
OUTSTAND	99.5% + rate of pick-up within 30 minutes of promised time.
ABVAVE	99.0% - 99.4% rate of pick-up within 30 minutes of promised time.
AVERAGE	90.0% - 98.9% rate of pick-up within 30 minutes of promised time.
BELOWAVE	Under 90.0% rate of pick-up within 30 minutes of promised time.

Sub-Criterion: FASTTRK	
Rating	**Definition**
OUTSTAND	Carrier has real-time tracking capabilities, and provides the shipper with communication links so that shipper can track a shipment without having to call the carrier.
ABVAVE	Carrier has real-time tracking capabilities, and shipper can call and receive immediate information on status of a shipment.
AVERAGE	Carrier has overnight batch-type tracking capabilities (i.e., status of shipment as of previous day is available).
BELOWAVE	Carrier has no automated tracking, and shippers' shipment status requests typically take several days.

Sub-Criterion: MKTVIEW	
OUTSTAND	The carrier is considered one of the market leaders in most or all major transport lanes of interest to the shipper, and has a reputation for innovative, mutually beneficial relationships with its customers.
ABVAVE	The carrier is considered at least a regional market leader in many major transport lanes of interest to the shipper, and has a reputation for innovative, mutually beneficial relationships with its customers.
AVERAGE	The carrier is a viable competitor in the particular transport lanes being evaluated by the shipper.
BELOWAVE	The carrier is a small player in the lanes of interest to the shipper, as well as the market as a whole, and does not have a strong reputation.

Figure 4.11 Examples Of Ratings Scale Descriptions For Selected Sub-Criteria

To illustrate our methodology, we consider the modal and carrier choice decision facing a hypothetical firm. This firm manufactures finished goods in a plant in the Far East and distributes these products to customers in Europe from one central distribution center located somewhere on the continent. The firm supplies make-to-stock products on demand. Thus, it must maintain inventory at the distribution center to fill orders immediately as customers place them, and it must maintain a safety stock or buffer inventory to cover the variability in demand over inventory replenishment lead time and the variability in replenishment lead time itself. For purposes of illustration, we will assume that although this firm ships and sells numerous products, all of its individual products have similar cost and demand characteristics. Thus, in this example, the firm will evaluate the best mode and carrier choice to ship a product family which represents an aggregation of similar individual end items. The question the firm must now answer is: which one of the available air and ocean carriers should it select to ship its products from its plant to its distribution center? To simplify the exposition, we assume that the firm is evaluating two air freight service providers and two ocean freight service providers; however, the basic methodology presented here can be applied to as complex a transportation flow or network as desired (e.g., intermodal alternatives and/or flows with transshipment locations).

Developing The AHP Model

A key first step in developing the AHP consists of identifying and then defining those factors to include as criteria. Transport cost clearly represents an important factor and is one of the criteria selected. Concerning the qualitative factors, recall that Figure 4.9 displays a set of decision criteria which are frequently used by shippers to help make transport carrier selections. Although we could have used all the criteria in this list, for illustrative purposes we assume that qualitative criteria 1, 2, 4, 5, and 6, as well as cost, represent the critical factors in determining the best transport alternative. In general, the complexity of an AHP model increases as the number of criteria increases. Therefore, in practice, one should carefully select only those criteria which significantly affect the modal and carrier choice decision.

To further clarify the meaning of each of the selected criteria, we define relevant sub-criteria, a modification of Approach 2. Figure 4.12 displays the selected criteria along with the definitions of the associated sub-criteria. Note that the number of sub-criteria employed to define a specific criterion varies. Customer Service (CUSTSER), a rather broad criterion, is comprised of five different sub-criteria, whereas Capacity (CAPACTY), a relatively straightforward criterion, requires only two.

Following Approach 2, we establish rating scales at the sub-criterion level. For simplicity we adopted a four-level ratings classification for each sub-criterion: Outstanding (OUTSTAND), above average (ABVAVE), average (AVERAGE), and below average (BELOWAVE). We require that each rating level for each sub-criterion be clearly defined and scaled to help standardize the evaluation process. Figure 4.11 displays the rating scales established for selected sub-criteria. Similar scales were constructed for each sub-criterion given in Figure 4.12.

The AHP Hierarchy

Figure 4.13 illustrates the complete AHP hierarchy for our transport mode and carrier selection problem. In total there are six levels in this hierarchy: the goal, the super-criteria, the criteria, the sub-criteria (for the qualitative sub-criteria only), the rating scales (for the qualitative sub-criteria only), and the carrier alternatives.

CUSTSER-Perceived Quality of *Customer Service*		
•	RESPTIM	*Response time* to the shipper's requests for transport service.
•	RELIABL	*Reliability* of transport services, in terms of meeting the promised time and day.
•	ADVNOTF	*Advanced notice* required for placing a shipment.
•	PROBRES	*Problem response* capability, both in terms of rapidly informing the shipper of in-transit problems, and responding quickly and effectively with alternative shipping solutions.
•	LOCSUPP	*Local support* capability, provided either at the shipper's site or through another type of customized local service.
TRACKTR- Shipment *Tracking* and *Tracing* Services		
•	FASTTRK	*Fast tracking and tracing* capabilities.
•	COVERAG	Seamless *coverage* from origin to destination without gaps.
•	ACCURAC	Relative *accuracy*.
EDI – *Electronic Data Interchange* (EDI) Capabilities		
•	EDISPAN	The *span* of EDI services available (e.g., billing, payment).
•	EDIEASE	The *ease* of interfacing/integrating with the carrier's EDI systems.
•	EDISHAR	The carrier's willingness to *share* software and/or hardware technology to facilitate effective EDI communications.
PARTNER- Potential to Develop Mutually Beneficial Long-Term *Partnership*		
•	MKTVIEW	The *market's view* (both shipper and carrier segments) of the specific carrier.
•	FINANCE	*Financial* strength of the carrier as measured by balance sheet and cash flow.
•	HISTORY	Relevant *historical* experience (if any) with the carrier.
CAPACTY- Cargo *Capacity* Limitations of the Carrier		
•	LIMITS	*Limitations* placed on regularly-available capacity (e.g., per shipment or per week).
•	PEAKPER	The ability to meet *peak period* capacity requirements (e.g., end of quarter or year).
TOTCOST- The *Total Costs* of shipping		
•	ANNCOST	The total *annual cost* comprised of freight and inventory carrying costs.
•	INVCOST	The total *inventory investment cost* of producing the inventory to fill and support the pipeline associated with a particular carrier.

Figure 4.12 Selected Criteria And Sub-Criteria For Hypothetical Example

The desire to represent the manner in which many shippers think about the modal choice decision motivated the inclusion of the two additional levels. One of these, the sub-criteria, was discussed previously. The other, namely the super-criteria, relates to the observation that many shippers classify criteria into the two generic categories of "quantitative or cost-related" and "qualitative or intangible", with weights either explicitly or implicitly assigned to each. Creating the "super-criteria" level enables a shipper to explicitly compare the relative importance of the cost vs. qualitative criteria in the mode and carrier choice decision. This leads to the development of a simple breakeven-type analysis (to be shown later) which displays the sensitivity of the carrier selection decision to the change in super-criteria weights.

For purposes of analysis and ease of exposition, we choose to view Figure 4.13 as containing two sub-hierarchies corresponding to the two super-criteria: (1) cost and (2) qualitative factors. We now discuss each of these components in turn.

Cost Analysis

The data in Tables 4.7 and 4.8 support the development of the cost component of the AHP analysis. (Alternatively, the cost methodology illustrated in these tables can stand on its own as a costing decision support tool.) Table 4.7 presents the inventory and freight costs of the four carriers in a spreadsheet-like format. In this example, we assume the firm would ship product to its European logistics center at the same shipment frequency or interval if it uses air carrier 1 or 2, or ocean carrier 1 (i.e., once every 7 days). However, if the firm uses ocean carrier 2, it would ship product only once every 28 days (e.g., ocean carrier 2 only offers a sailing once every four weeks). This results in both a larger average shipment quantity and larger average cycle stock inventory if the firm uses ocean carrier 2 rather than ocean carrier 1, or either of the air carriers.

In practice, there exists an extensive, well known body of research on different ways to establish the relationship between shipment frequency and order size (e.g., the EOQ model, the fixed order period model, and so on). Additionally, there are a variety of heuristic methods used in actual practice. For illustrative purposes, and following a standard practice of many shippers, we assume a fixed shipment frequency, and then derive the shipment size. Table 4.8 provides a summary comparison of both the pipeline investment and annual costs by carrier. The analysis indicates that either of the two ocean carriers will generate substantial annual cost savings for the shipper relative to either air freight carrier (column 5, Table 4.8). The annual freight savings associated with ocean transport (column 4,

Table 4.7 Development Of Inventory And Freight Costs By Carrier

(1) Mode and Carrier	(2) Annual Demand Forecast (units)	(3) Transit Times: Door-to-Door (days)	(4) Average In-Transit Inventory in Pipeline (units)	(5) Frequency of Shipment Arrivals: (i.e., once every__) (days)	(6) Shipment Quantity (units)	(7) Average Cycle Stock Inventory at Logistics Center (units)
AIR CARRIER 1	23,000	7	441	7	442	221
AIR CARRIER 2	23,000	7	441	7	442	221
OCEAN CARRIER 1	23,000	28	1,764	7	442	221
OCEAN CARRIER 2	23,000	35	2,205	28	1,764	882

(8) Average Safety Stock Required at Logistics Center When Replenished by Air or Ocean Carrier (units)	(9) Total Inventory In-Transit and at Logistics Center (units)	(10) Product Cost ($/unit)	(11) Total Inventory In-Transit and at Logistics Center (@ cost)	(12) Freight Cost Door-to-Door ($/unit)	(13) Total Freight Cost Per Year ($)
885	1,547	$950	$1,469,650	$55	$1,265,000
885	1,547	$950	$1,469,650	$52	$1,196,000
1,770	3,755	$950	$3,567,250	$13	$299,000
1,979	5,066	$950	$4,812,700	$11	$253,000

Table 4.7 Development Of Inventory And Freight Costs By Carrier (continued)

Notes for Table 4.7

- Columns 2, 3, 5, 10 and 12 are inputs

Column Formulas:

(4) = (2) x [(3)/365]

(6) = (2) x [(5)/365] . . . note: calculations for the two air carriers and ocean carrier 1 were rounded up to 442 to create an integer result in column (7)

(7) = (1/2) x (6)

(8) [Note: All safety stock assumptions for this example are illustrative.] For Air Carrier 1 and Air Carrier 2, safety stock was assumed equal to two weeks of demand; i.e., (2) x (2/52).
For ocean: [Square root of (ocean trans time/air trans time)] x air safety stock e.g. SQRT (28/7) x 885 = 1,770; for illustrative purposes this calculation was loosely patterned after the square root law, see Zinn et al. (1989)

(9) = (4) + (7) + (8)

(11) = (9) x (10)

(13) = (12) x (2)

Table 4.8 Summary of Inventory Investment And Annual Costs By Carrier

(1) Mode and Carrier	(2) Inventory Investment Cost ($)	(3) Annual Carrying Costs ($)	(4) Annual Freight Costs ($)	(5) Total Annual Costs ($)
Air Carrier 1	1,469,650	293,930	1,265,000	1,558,930
Air Carrier 2	1,469,650	293,930	1,196,000	1,489,930
Ocean Carrier 1	3,567,250	713,450	299,000	1,012,450
Ocean Carrier 2	4,812,700	962,540	253,000	1,215,540

Column Formulas:

(2) is taken from (11), Table 4.7

(3) = .20 x (2); i.e., we assume a 20% annual inventory carrying cost interest rate

(4) = is taken from (13), Table 4.7

(5) = (3) + (4)

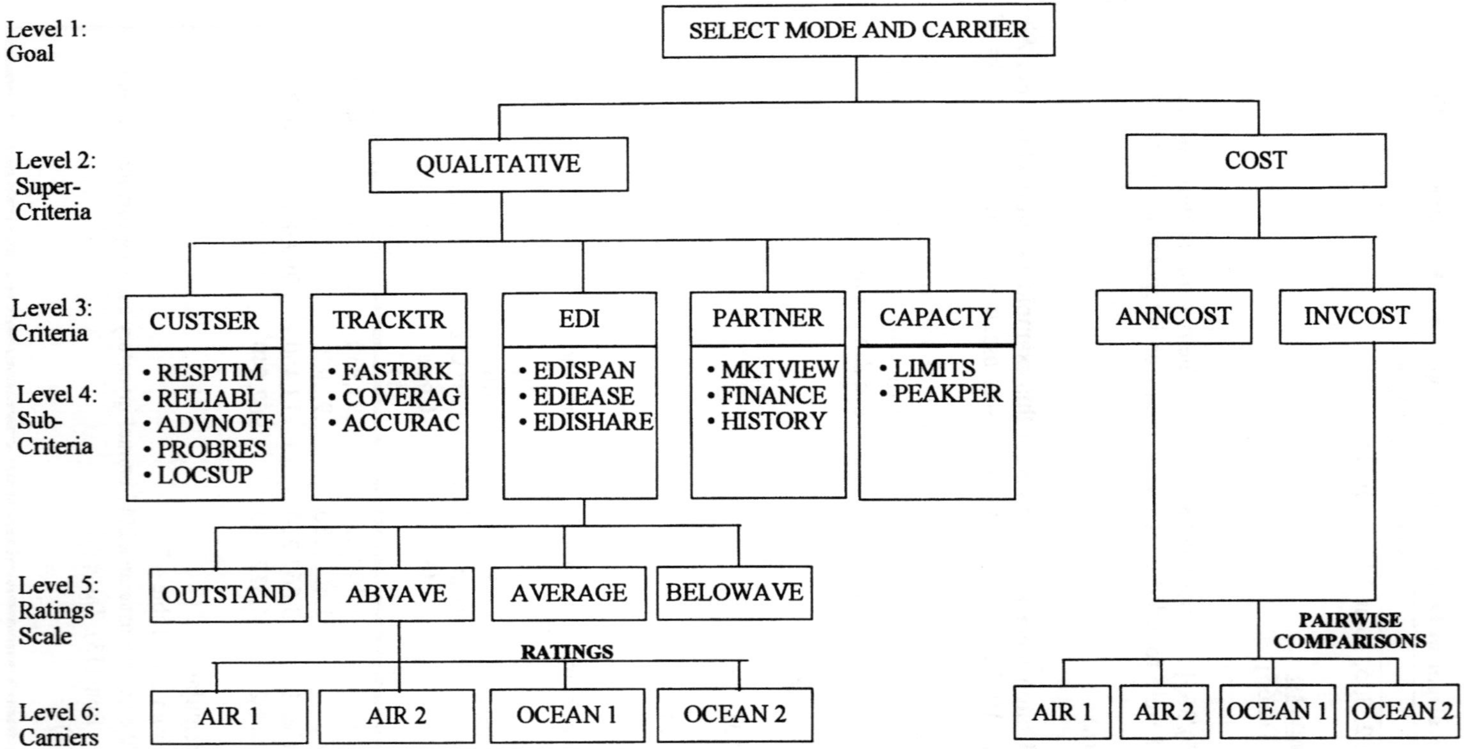

Figure 4.13 AHP Hierarchy For Transport Mode and Carrier Selection

Table 4.8) far outweigh the higher annual inventory carrying costs of this mode (column 3, Table 4.8). However, to generate these annual cost savings, the shipper must make a much larger investment in inventory to support the transport pipeline (column 2, Table 4.8). Note that using either ocean carrier will require greater "in-transit" inventory (column 4, Table 4.7) and greater safety stock (column 8, Table 4.7) relative to either air carrier. Thus, the shipper must trade-off annual cost savings versus a larger investment in inventory (i.e., working capital). The AHP is an ideal vehicle to operationalize this trade-off.

Three sets of judgments are required: 1) pairwise comparisons of the four carriers relative to ANNCOST; 2) pairwise comparisons of the four carriers relative to INVCOST; and 3) a pairwise comparison between investment costs and annual costs to weigh the results from the first two analyses (Table 4.9). The summary results found in Table 4.8 facilitate the development of the judgments in the ANNCOST and INVCOST matrices. Note that 1's appear along the main diagonal in each pairwise comparison matrix since any item is equally preferred to itself. Also, the pairwise comparisons below the main diagonal are simply the reciprocals of the corresponding entries above the main diagonal. As expected, the pairwise comparisons indicate that our shipper prefers the ocean carriers to the air carriers with respect to ANNCOST, while displaying the opposite preference with respect to INVCOST. The strength of these preferences, however, results from our shipper's assessment of the differences in the magnitudes between the various costs shown in Table 4.7. The final carrier weights with respect to COST (bottom of Table 4.9) are based on a 60-40 split of total weight between the annual cost and investment cost sub-criteria. Under this assumption, ocean carrier 1 is the most favored alternative from the perspective of total costs.[12]

Qualitative Analysis

Pairwise comparisons must be collected and processed for the rating scales[13], sub-criteria, and criteria. These data are presented in Table 4.10. Using the same weighted averaging approach previously employed, we can compute the weight associated with each of the four rating levels for each criterion. The weights for the selected ratings can be added for a total project score and then renormalized to sum to one.

---

[12] All AHP analyses were accomplished using the microcomputer program Expert Choice (Foreman et. al., 1990).

[13] To further simplify the analysis, we develop only one set of rating scale judgements that is used for all sub-criteria. That is, the relative relationship between OUTSTAND, ABVAVE, AVERAGE, AND BELOWAVE is the same for all sub-criteria (see Liberatore, Nydick and Sanchez for additional discussion of this approach).

Table 4.9 Pairwise Comparison Matrices And Weights For Cost Analysis

ANNCOST Comparison Matrix and Weights

	Air Carrier 1	Air Carrier 2	Ocean Carrier 1	Ocean Carrier 2	Weights
Air Carrier 1	1	1/2	1/7	1/8	0.052
Air Carrier 2	2	1	1/6	1/5	0.084
Ocean Carrier 1	7	6	1	2	0.511
Ocean Carrier 2	8	5	1/2	1	0.354
					1.000

INVCOST Comparison Matrix and Weights

	Air Carrier 1	Air Carrier 2	Ocean Carrier 1	Ocean Carrier 2	Weights
Air Carrier 1	1	1	7	8	0.439
Air Carrier 2	1	1	7	8	0.439
Ocean Carrier 1	1/7	1/7	1	2	0.073
Ocean Carrier 2	1/8	1/8	1/2	1	0.048
					1.000

Cost Criteria Comparison Matrix and Weights

	ANNCOST	INVCOST	Weights
ANNCOST	1	3/2	0.600
INVCOST	2/3	1	0.400
			1.000

Carrier Weights with Respect to Cost

	ANNUAL 0.6	INVEST 0.4	Weights
Air Carrier 1	0.052	0.439	**0.207***
Air Carrier 2	0.084	0.439	**0.226**
Ocean Carrier 1	0.511	0.073	**0.336**
Ocean Carrier 2	0.354	0.073	**0.231**

* This weight is computed as follows: (0.600)*(0.052) + (0.400)*(0.439) = 0.207

Table 4.10 Pairwise Comparison Matrices And Weights For Analysis Of Qualitative Evaluation Factors

RATINGS SCALE

	OUTSTAND	ABVAVE	AVERAGE	BELOWAVE	Weights
OUTSTAND	1	2	3	5	0.473
ABVAVE	1/2	1	2	4	0.284
AVERAGE	1/3	1/2	1	3	0.170
BELOWAVE	1/5	1/4	1/3	1	0.073
					1.000

CUSTSERV Sub-Criteria

	RESPTIM	RELIABL	ADVNOTF	PROBRES	LOCSUP	Weights
RESPTIM	1	1/2	3	1/3	2	0.174
RELIABL	2	1	2	1/2	3	0.243
ADVNOTF	1/3	1/2	1	1/3	1	0.097
PROBRES	3	2	3	1	4	0.399
LOCSUP	1/2	1/3	1	1/4	1	0.088
						1.000

TRACTTR Sub-Criteria

	FASTTRK	COVERAG	ACCURAC	Weights
FASTTRK	1	3	1/2	0.309
COVERAG	1/3	1	1/5	0.109
ACCURAC	2	5	1	0.582
				1.000

EDI Sub-Criteria

	EDISPAN	EDIEASE	EDISHARE	Weights
EDISPAN	1	2	7	0.615
EDIEASE	1/2	1	3	0.292
EDISHARE	1/7	1/3	1	0.093
				1.000

PARTNER Sub-Criteria

	MKTVIEW	FINANCE	HISTORY	Weights
MKTVIEW	1	1/3	1/4	0.126
FINANCE	3	1	1	0.416
HISTORY	4	1	1	0.458
				1.000

CAPACTY Sub-Criteria

	LIMITS	PEAKPER	Weights
LIMITS	1	3	0.750
PEAKPER	1/3	1	0.250
			1.000

QUALITATIVE Criteria

	CUSTSER	TRACKTR	EDI	PARTNER	CAPACITY	Weights
CUSTSER	1	3	3	2	4	0.409
TRACKTR	1/3	1	1/2	1/2	1	0.103
EDI	1/3	2	1	2	3	0.229
PARTNER	1/2	2	1/2	2	3	0.170
CAPACTY	1/4	1	1/3	1/3	1	0.089
						1.000

Table 4.11 Carrier Rating Spreadsheet

	CARRIERS				
	Air 1	Air 2	Ocean 1	Ocean 2	Weights
CUSTSER					
RESPTIM					
OUTSTAND	0.000	0.034	0.000	0.000	0.034
ABVAVE	0.020	0.000	0.000	0.000	0.020
AVERAGE	0.000	0.000	0.012	0.012	0.012
BELOWAVE	0.000	0.000	0.000	0.000	0.005
RELIABL					
OUTSTAND	0.000	0.000	0.000	0.000	0.047
AVBAVE	0.028	0.028	0.000	0.000	0.028
AVERAGE	0.000	0.000	0.017	0.000	0.017
BELOWAVE	0.000	0.000	0.000	0.007	0.007
ADVNOTF					
OUTSTAND	0.000	0.019	0.000	0.000	0.019
ABVAVE	0.011	0.000	0.011	0.000	0.011
AVERAGE	0.000	0.000	0.000	0.007	0.007
BELOWAVE	0.000	0.000	0.000	0.000	0.003
PROBRES					
OUTSTAND	0.000	0.000	0.000	0.000	0.077
ABVAVE	0.046	0.000	0.046	0.000	0.046
AVERAGE	0.000	0.028	0.000	0.028	0.028
BELOWAVE	0.000	0.000	0.000	0.000	0.012
LOCSUPP					
OUTSTAND	0.017	0.000	0.000	0.000	0.017
ABVAVE	0.000	0.010	0.000	0.000	0.010
AVERAGE	0.000	0.000	0.006	0.000	0.006
BELOWAVE	0.000	0.000	0.000	0.003	0.003
TRACKTR					
FASTTRK					
OUTSTAND	0.000	0.015	0.000	0.000	0.015
ABVAVE	0.009	0.000	0.000	0.000	0.009
AVERAGE	0.000	0.000	0.000	0.005	0.005
BELOWAVE	0.000	0.000	0.002	0.000	0.002
COVERAG					
OUTSTAND	0.000	0.005	0.000	0.000	0.005
ABVAVE	0.003	0.000	0.000	0.000	0.003
AVERAGE	0.000	0.000	0.002	0.002	0.002
BELOWAVE	0.000	0.000	0.000	0.000	0.001
ACCURAC					
OUTSTAND	0.028	0.028	0.000	0.000	0.028
ABVAVE	0.000	0.000	0.000	0.017	0.017
AVERAGE	0.000	0.000	0.010	0.000	0.010
BELOWAVE	0.000	0.000	0.000	0.000	0.004

Table 4.11 Carrier Rating Spreadsheet (continued)

	CARRIERS				
	Air 1	**Air 2**	**Ocean 1**	**Ocean 2**	**Weights**
EDI					
EDISPAN					
OUTSTAND	0.000	0.067	0.000	0.000	0.067
ABVAVE	0.040	0.000	0.000	0.040	0.040
AVERAGE	0.000	0.000	0.024	0.000	0.024
BELOWAVE	0.000	0.000	0.000	0.000	0.010
EDIEASE					
OUTSTAND	0.000	0.000	0.000	0.000	0.032
ABVAVE	0.000	0.019	0.000	0.000	0.019
AVERAGE	0.011	0.000	0.000	0.000	0.011
BELOWAVE	0.000	0.000	0.005	0.005	0.005
EDISHARE					
OUTSTAND	0.000	0.000	0.000	0.000	0.010
ABVAVE	0.006	0.006	0.000	0.000	0.006
AVERAGE	0.000	0.000	0.004	0.004	0.004
BELOWAVE	0.000	0.000	0.000	0.000	0.002
PARTNER					
MKTVIEW					
OUTSTAND	0.000	0.010	0.000	0.000	0.010
ABVAVE	0.006	0.000	0.000	0.006	0.006
AVERAGE	0.000	0.000	0.004	0.000	0.004
BELOWAVE	0.000	0.000	0.000	0.000	0.002
FINANCE					
OUTSTAND	0.000	0.000	0.000	0.000	0.033
ABVAVE	0.000	0.020	0.000	0.000	0.020
AVERAGE	0.012	0.000	0.012	0.012	0.012
BELOWAVE	0.000	0.000	0.000	0.000	0.005
HISTORY					
OUTSTAND	0.000	0.000	0.000	0.000	0.037
ABVAVE	0.022	0.022	0.000	0.022	0.022
AVERAGE	0.000	0.000	0.013	0.000	0.013
BELOWAVE	0.000	0.000	0.000	0.000	0.006
CAPACTY					
LIMITS					
OUTSTAND	0.000	0.000	0.031	0.000	0.031
ABVAVE	0.000	0.019	0.000	0.019	0.019
AVERAGE	0.011	0.000	0.000	0.000	0.011
BELOWAVE	0.000	0.000	0.000	0.000	0.005
PEAKPER					
OUTSTAND	0.000	0.000	0.010	0.000	0.010
ABVAVE	0.000	0.006	0.000	0.006	0.006
AVERAGE	0.004	0.000	0.000	0.000	0.004
BELOWAVE	0.000	0.000	0.000	0.000	0.002
RAW SCORE	0.277	0.337	0.210	0.195	
RENORM SCORE*	**0.272**	**0.331**	**0.207**	**0.191**	

* Raw score/sum of raw scores; necessary so that scores sum to one.

The project rating spreadsheet is given as Table 4.11. The last column of Table 4.11 displays the weights for each rating level for each sub-criterion. Based on the hypothetical ratings shown we see that air carrier 2 is the highest ranking carrier with respect to the qualitative criteria (see the last row of Table 4.11).

Combining The Cost And Qualitative Analyses

Using the underlined carrier weights for the cost (bottom of Table 4.9) and qualitative (bottom of Table 4.11) super-criteria, we can summarize the carrier selection decision as shown in Figure 4.14. By varying the cost super-criterion weight between 0 and 1, we can see the ranges over which each specific carrier is preferred. Each line in Figure 4.14 shows the final weight of a given carrier as a function of the cost super-criterion weight (remember that the qualitative super-criterion weight equals one minus the cost super-criterion weight). In particular, for a cost weight of .530 or less, air carrier 2 is preferred; otherwise ocean carrier 1 should be selected. For example, assume that a shipper thought that total costs should receive a 75% weight and qualitative factors about a 25% weight. In this case, the shipper clearly should use ocean carrier 1. Similarly, suppose a shipper could not decide on the weight to assign to total cost, however, he or she felt comfortable that the weight for cost ranged between 60% and 80%. Then the shipper again could conclude that ocean carrier 1 represented the best choice. In cases where a shipper finds that his or her weight assignment between the super-criteria approaches the breakeven point (.53 in this example), he or she may want to re-examine some previously determined pairwise comparisons or ratings for additional insights. For example, one could perform sensitivity analysis to observe the impact of the weights assigned to the investment cost vs. annual cost (Table 4.9), on the ultimate breakeven chart (Figure 4.14). Figure 4.14 also shows that air carrier 1 and ocean carrier 2 should never be used (both are dominated by the other carrier in their respective modes). As noted though, the sensitivity of the final decision can be found with respect to other factors as well, leading to somewhat different slopes for the lines shown in Figure 4.14.

Summary And Conclusions

The AHP approach for carrier and mode selection presented in this hypothetical case study offers several important advantages. First, in line with the practice of many shippers, carrier alternatives are evaluated by explicitly trading off the results obtained from the distinct analyses of costs and qualitative factors. Second, the judgements required for the cost sub-hierarchy are based on the results of a complete logistics costing model rather than a consideration of carrier rates alone. Third, each qualitative criterion has a well-defined rating scale that simplifies and clarifies this portion of the evaluation process. Fourth, what-if and sensitivity analyses can be accomplished easily at any level of the analysis using available AHP and spreadsheet software.

4.3.2 Developing Product Families For Mode Choice

For illustrative purposes, the presentation in section 4.3.1 made the assumption that a firm was making a mode choice decision for a product family of similar end items. In fact, creating the proper product families represents a small, but important element of the analytic decision support required for making effective mode choice decisions. Traditionally, one develops product families for hierarchical production and distribution planning based on such characteristics as the similarities of individual end items in terms of their supply sources, weight, manufacturing and perhaps distribution characteristics. For production planning purposes, for example, a product family is typically composed of those end items that can all be produced on the same production line or lines, at similar rates and costs per time period without major changeovers in between the production of one end item to the next. From a distribution perspective, similarity of handling requirements can also form the basis of grouping end items into product families, in addition to those characteristics previously mentioned.

For mode choice planning purposes, one must use somewhat different factors as the basis for grouping end items. The principal determinants to consider in developing product families for mode choice analyses include: (1) similarity of handling requirements and (2) the product value (i.e., cost) of an end item. Handling requirements refer to such distribution shipping factors as temperature control requirements, hazardous material restrictions, and clearly must be considered. Product value, however, plays an extremely important role in creating

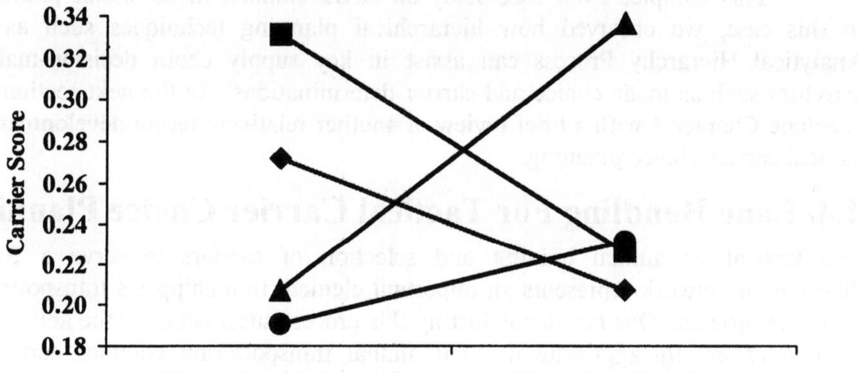

Example of computation of a carrier's score

- Suppose the "cost weight" = 0.6
- Then the "qualitative weight" = 1 -.06 = 0.4
- Air Carrier 1's score would then be computed as follows:

(.6 x .207) + (.4 x .272) = .233

where: .207 = Air Carrier 1's weight with respect to cost [bottom of Table 4.9]

.272 = Air Carrier 1's weight with respect to qualitative factors [bottom of Table 4.11]

Figure 4.14 Carrier Score Vs. Cost Weight

mode choice product families. This results from the relationship between product cost and inventory carrying cost. In trade-offs between modes that have significant differences in transit time (e.g., air vs. ocean), if an end item has a high product cost, the inventory carrying costs associated with one mode compared to another can vary dramatically. This is particularly the case when one considers that transit time differences between modes translate into not only differences in in-transit inventory carrying costs, but also differences in safety stock requirements at receiving locations (e.g., the example in section 4.3.1). In contrast, the inventory carrying costs associated with different modes will not vary so significantly when a product has a relatively low cost. In general, between any origin-destination pair on a network where a planner must make a modal choice decision, it is often possible to group a wide range of individual end items into two or perhaps at most three or so product families. Assuming that all end items at an origin have similar general shipping characteristics (which is often the case), one simply needs to create groupings based around product cost. It is not an unusual situation in some industries to find that one group of end items should be shipped between an origin-destination pair via air, while a second group should be shipped via surface mode. In the computer industry for example, a firm may find that it should ship pc's and higher value products via air between a plant and a DC, while it can most economically ship lower value peripheral products via surface.

This completes our case study on tactical/annual mode choice planning. In this case, we observed how hierarchical planning techniques such as the Analytical Hierarchy Process can assist in key supply chain decision-making activities such as mode choice and carrier determinations. In the next section, we conclude Chapter 4 with a brief review of another relatively recent development in tactical carrier choice planning.

4.4 Lane Bundling For Tactical Carrier Choice Planning

The tactical or annual bidding and selection of carriers to serve a firm's distribution network represents an important element in a shipper's transportation planning process. One key to conducting this process successfully is the ability of a firm to *effectively* aggregate its total annual transportation volumes across its network to present requests for quotes to its potential transport carriers. Simply summing up total volumes on all lanes or groups of lanes across a network does not necessarily assure that a firm will leverage its annual transportation volumes most effectively in the bidding process. Rather, planners should employ a decision support methodology that guides the aggregation of volumes, lanes and groups of lanes such that the ultimate freight volume presented to carriers for bids facilitates an economical and effective bid for both parties. Thus, the annual or tactical[14] carrier bidding and selection process has a hierarchical planning element; namely, the process of aggregating planned shipments and transport lanes. In this section, we briefly review one relatively recent optimization model-based methodology

[14] We refer to the carrier bidding and selection process as annual or tactical in recognition that some firms bid their transport volumes on an annual basis, while others choose to solicit bids on a less frequent basis (e.g., once every two years).

implemented and used in private industry to facilitate the transportation bidding process.

4.4.1 Case Review: A Transportation Bidding Software System

In late 1996, a transportation bidding software system now known as OptiBid became available on the commercial market.[15] This section briefly describes the carrier selection process facilitated by OptiBid in order to illustrate the types of activities involved in this process. Additionally, this review will further illustrate the hierarchical characteristics of the carrier selection process.

The objective of the OptiBid software system and methodology is to "determine the optimal assignment of carriers to lanes in a shipper's transportation network".[16] The methodology structured by OptiBid attempts to develop this optimal assignment by considering the transportation bid from the perspective of both the shipper requesting the bids, as well as the carriers from whom a shipper chooses to solicit bids. This methodology employs the premise that by evaluating both its own transport volume requirements and the capabilities of its existing and potential carriers, a shipper can facilitate more efficient, economical bids from both parties' perspectives.

OptiBid presumes that a shipper will engage in a three step carrier selection process consisting of:

1. Bid preparation and network analysis (performed by the shipper)
2. Carrier analysis (performed by the carriers), and
3. Bid analysis (performed by the shipper)

Briefly, in step 1, the shipper uses historical data and/or forecasts, as well as other pertinent network data to develop transport volumes for carriers to bid upon. In step 2, the carriers evaluate the potential transport volumes and service requirements presented by the shipper and carriers then submit specific bids for parts or all of the business. Finally in step 3, the shipper evaluates the collective set of bids received and determines which carriers to use for different transport lanes and volumes. The reader interested in specific details of how OptiBid functions is referred to the literature available on this software. However, we now highlight certain elements of the overall process that have similarities to other hierarchical planning approaches previously reviewed.

In the bid preparation and network analysis step, shippers must aggregate the individual shipments that occur across their networks into larger groups of shipments and groups of lanes to offer to carriers. Shippers use OptiBid to create these groupings based upon such characteristics as:

---

[15] The SABRE Group, headquartered in Dallas, Texas offers OptiBid. The description of this software presented in this section is based entirely upon the SABRE Group's documentation on this software. It should be noted, however, that the author's intent here is simply to briefly describe how software such as OptiBid can fit into a hierarchical transportation planning approach. This description of OptiBid does not represent an endorsement by the author of OptiBid relative to other competitive products.

[16] As described by the SABRE Group in its literature on OptiBid.

- the proximity of the origins of shipments
- the proximity of the destinations of shipments
- the length (i.e., distance) of shipments
- the type of transport equipment required for shipments
- the transit time requirements of shipments, and
- the type of shipment (inbound to shipper or outbound to customer)

The above list represents just a sample of the potential criteria that one can employ to generate shipment groupings. Note that this grouping activity required to facilitate tactical (annual) transportation bidding parallels the "grouping" work required to create product families for the tactical production planning level of hierarchical production planning systems.

A second step in the bid preparation and network analysis process consists of determining which segments of potential transportation business to offer to different potential carriers. In this step, the shipper can choose to tailor its requests for bids to subsets of carriers for subsets of the total transportation volume. The shipper makes these determinations based upon an evaluation of the potential carriers' capabilities and network infrastructure. In the vernacular of hierarchical production planning; therefore, the shipper needs to "decompose" the planning problem (i.e., the solicitation of transportation bids) into a series of subproblems in order to facilitate an efficient bidding process.

The OptiBid-based methodology advocates a similar decomposition of the bidding problem from the carriers' perspectives in step 2, as the following example illustrates. The price that a carrier must charge for its services to generate an acceptable profit is significantly affected by the level of equipment utilization that a carrier achieves. For example, if a shipper wants a carrier to move a load from point A to point B, the price that the carrier must charge to assure a profit on this business will be affected by whether or not the carrier can: (1) find another load to move to point A (before picking up the load at point A), and (2) find another load to move from point B onward (after moving the shipper's load from A to B). If the carrier can essentially utilize its equipment (e.g., tractor and trailer) *continuously* before and after the load from A to B, it can charge a lower rate for the A to B service than if the carrier must travel empty for a significant distance either before picking up the load at A or after dropping the load at B. OptiBid addresses the economics[17] of carrier costs illustrated in this example by allowing the carrier to submit multiple bids. Specifically, a carrier can submit at least two different price bids for the service from A to B in the above example. A carrier can submit one price just for the service from A to B. In addition, the carrier can submit a second "conditional" price bid on the service from A to B. This second (and presumably lower) price bid could be based on the condition that the shipper also offer the carrier either a move inbound to point A and/or a move outbound from point B. The second lower price on the A to B move would be effective only if the carrier received this additional inbound and/or outbound business from the shipper. Thus,

[17] The costs illustrated in this example are often referred to as economies of scope.

by facilitating both single lane and multiple lane bids from individual carriers with conditional bids allowed, the OptiBid methodology attempts to minimize a shipper's costs, subject to service requirements, based on both the shipper's and its suppliers needs and capabilities.

In step 3, the shipper evaluates the individual carrier bids received and selects the carrier or carriers that it will designate as its transport service provider for particular lanes and/or groups of lanes. OptiBid employs what the firm's literature describes as a "combinatorial optimization problem" to make the determinations of which carriers should serve particular lanes. This optimization model evaluates both the conditional and "non-conditional' bids presented by potential carriers, as well as a series of service related or qualitative factors applied by the shipper (if the shipper chooses to include qualitative factors). The service related factors that a shipper can include in the optimization model are such factors as on-time performance, EDI capabilities, damage rates, and so on. These service factors are entered into the optimization model as penalty factors. For example, a shipper can assign a dollar penalty factor (e.g., $50 per load) to the bids of carriers with historical on-time performance records below some required measure (e.g., 90%). Thus, the OptiBid methodology is similar to the AHP-based mode selection methodology discussed in section 4.3 in that it also recognizes the importance of qualitative and non-direct freight charge costs in the carrier selection process. Based upon the recommendations generated by optimization model runs and a shipper's overall judgements, step 3 finally culminates with the selection of carriers by the shipper.

4.4.2 Operational Feedback In The Carrier Selection Process

Similar to the feedback required in hierarchical production planning processes, feedback from the operational level to the tactical level represents a critical element of the transportation selection process. A tactical planning process such as the one described in section 4.4.1 results in the selection of a set of carriers to provide transport services for a firm's network. At the operational planning and execution level, a shipper then actually tenders loads to carriers for individual shipments, monitors the performance of carriers on individual shipments, monitors the availability of carriers against expected capacity (i.e., the availability promised during the tactical bidding process) and so on. The feedback required from the operational level to the tactical level in the carrier selection process has at least two major components. The first component consists of the aforementioned operational monitoring that must occur. For example, suppose that a designated carrier for a particular lane does not consistently have the right equipment available, or does not deliver expected on-time performance, or has a higher damage/loss rate than expected. This information must consistently flow back to a firm's tactical planners so that they can assist operational personnel in resolving any carrier performance issues. At the same time, this feedback becomes an important input for future tactical carrier selection processes. The second feedback component that must flow from the operational to tactical planning levels concerns the shippers performance vis-à-vis commitments that the shipper made to carriers in the carrier bid selection process. For example, suppose that the shipper guaranteed a minimum level of transportation volume to a particular carrier on a lane in return for a favorable freight rate as part of the bidding/negotiation process. Suppose further that the actual loads tendered to this particular carrier have not met the guaranteed minimum level during the first portion of the planning horizon. This represents the type of information that must be fed back to the tactical planning level as soon as possible to determine how to address the situation.

We will not belabor with additional examples the importance of an appropriate flow of feedback from the operational level to the tactical level. However, this two-way communication of information remains just as important to effective distribution and carrier operations as to effective hierarchical production planning processes and operations.

4.5 Summary

In this chapter, we have reviewed a wide range of distribution and transportation planning techniques for strategic, tactical and operational planning. As noted in the beginning of the chapter, the type of planning activities discussed here generally occur independently of, or "loosely integrated" with manufacturing planning. Further, in most cases, the separation of the planning activities documented in this chapter from manufacturing planning activities represents a reasonable decomposition of a firm's overall strategic, tactical and operational planning activities. Naturally, as one engages in these distribution and transportation planning exercises, it remains critical to always ask the question of whether manufacturing planning can be separated from a particular distribution and transportation planning problem without undue risk of developing significantly suboptimal results.

Chapter 5

Warehouse Design And Management

Maintaining effective and efficient warehouse operations necessitates integrated supply chain planning across multiple functional areas just as other operational components of the supply chain similarly depend upon integrated planning approaches. Further, integrated warehouse design, planning and management requires the use of a hierarchical paradigm, just as is required for the planning of other key supply chain components such as production.

In this chapter, we will explore how warehouse planning and operations fits into an overall hierarchical production and distribution planning scheme for a firm's supply chain. Section 5.1 begins with an overview of the warehouse planning process from strategic site selection down to operations level policies and procedures. In the next section, we consider how such activities as forecasting, production, inventory, and distribution network planning impact warehouse operations. Additionally, we review some practical decision support modeling techniques that firms can employ to link different planning efforts across the supply chain into warehouse planning and operations. Finally, section 5.3 concludes this chapter with a discussion of an integrated hierarchical set of models for warehouse planning.

5.1 The Warehouse Planning Process

The warehouse planning process begins at the network-wide strategic planning level. At this level, a firm must determine how warehouse operations fit into its overall strategic plan, and in particular, *what is the mission* of the warehouses or distribution centers on its network.[1] For starters, a firm must determine if it really requires stand-alone warehouse operations, or for example, if it wishes to employ a strategy of shipping directly from its plants to its customers. Throughout the 1990's, the "plant direct" strategy has attracted more and more interest in various industries (e.g., the personal computer manufacturing industry) as some firms have viewed this strategy as a means to minimize their overall inventory investment, and/or to compress overall order cycle times. Once a firm decides that it will in fact employ distribution centers to inventory product, perform customer fulfillment operations and so on, it must then consider such factors as overall network design and the appropriate total number of, and locations for, individual warehouses. Figure 5.1 provides a high level overview of this hierarchical planning process that begins at the strategic level, while Figure 5.2 illustrates some of the key questions and issues that planners must evaluate at each level. We will now consider these in more detail.

[1] We will use warehouse and distribution center as interchangeable synonyms in this chapter.

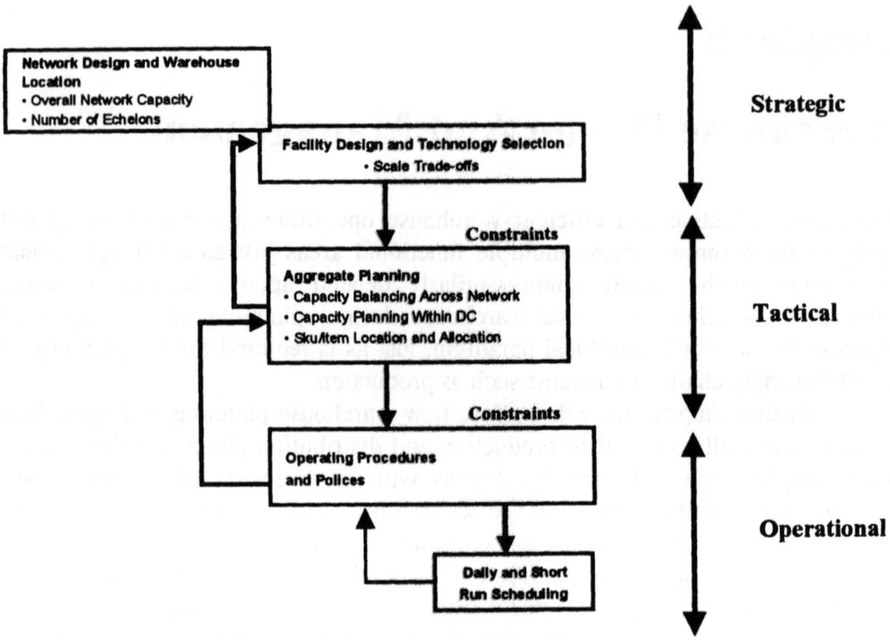

Figure 5.1 Hierarchical Warehouse Planning

A first step in the planning process consists of determining the mission of the overall warehouse network and the individual locations that will make up the network. Note that not all warehouses on a network will necessarily have the same mission or play the same role. For example, the practice of "cross-docking" where products flow inbound to a distribution center and are received and then immediately shipped out to the next "ship to" location on the supply chain (e.g., the customer's warehouse or retail location) has garnered much interest in recent years. Actual implementations of "cross-docking" warehouse strategies abound in such industries as retail merchandising, groceries, consumer products and others. A decision to employ "cross-docking" as a major warehousing strategy will significantly affect virtually all aspects of the warehouse planning process from the strategic level down to the operational level. Further, the question of whether "cross-docking" will represent the dominant warehousing approach of a firm, or perhaps just one of several key "product flow" strategies will affect the entire planning process. A firm may decide upon a segmentation approach where it employs cross-docking only for certain customer channels and/or certain product lines. It is not unusual to find large scale warehouse networks that employ a wide mix of product flow and product inventorying strategies, and that have a broad mix of distribution facilities with differing capabilities supporting these strategies (e.g., Wal*Mart).

The number of warehouse echelons to establish represents another common strategic network design question that heavily influences the mission of individual warehouses. For example, a firm must decide whether it will operate a single echelon network in which every warehouse will receive shipments of all products directly from all plants, or alternatively does the firm want to operate a multi-echelon warehouse network where one or more first echelon, central warehouses receive products from plants and then redistribute some or all products to second echelon regional/local warehouses. The number of choices regarding the echelon and mission strategy is quite complex and has ramifications throughout all planning and implementation levels. Finally, another important strategic decision concerns the question of whether a firm chooses to operate its own facilities or to outsource some or all of its warehouse operations to one or more third party providers. Although many firms still choose to operate their own networks, the use of third party warehouse firms has grown dramatically during the 1990's and prospects for future growth in this industry remain very strong (see e.g., Logistics Magazine, 1999). Thus, a firm must determine whether it will rely on its own expertise, on that of third parties or a combination of both to run its warehousing network.

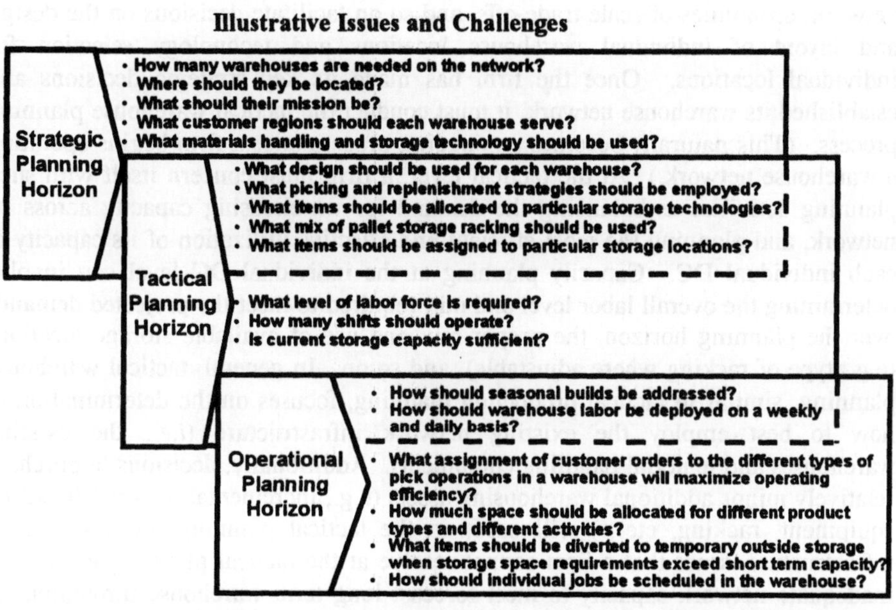

Figure 5.2 How Warehousing Decisions Fit Into A Planning Hierarchy

In determining the appropriate number of, and locations for individual warehouses, a firm's planners must evaluate how these locations will fit into its overall network. Thus, the locations and capabilities of the plants and other suppliers on the network, as well as the locations of the customers (or "ship-to" locations) of the firm's warehouses all heavily influence the warehouse network design problem. In Chapters 2 and 4, we previously observed how such planning techniques as mathematical optimization can provide critical decision support to the actual warehouse network location problem.

As Figure 5.1 illustrates, total network warehouse capacity requirements and economies of scale trade-offs are two additional key determinants of the interrelated decisions on network design, facility design and warehouse technology selection. The network capacity decision includes both storage capacity requirements and throughput capacity requirements. Capacity requirements, as well as the number of locations operated, can significantly impact the potential economies of scale achievable on the network and the level of investment in sophisticated warehousing technology that a firm can undertake. In general, the higher the throughput volume of an individual warehouse, the higher the level of investment in materials handling and automation technologies that a firm can make in a particular facility. Naturally, the more warehouses that a firm operates on a network for a given level of throughput, the lower the volume per facility; and therefore, the lower the investment per facility that a firm can justify. In selecting the best number, mix and locations for the warehouses on a network, a firm must carefully evaluate these trade-offs. Again, all of these decisions made at the strategic level will affect tactical and operational warehouse planning also.

Strategic decisions regarding the overall mission of the warehouse network, economies-of-scale trade-offs, and so on facilitate decisions on the design and layout of individual warehouse locations and technology selection for individual locations. Once the firm has made its key strategic decisions and established its warehouse network, it must conduct the tactical warehouse planning process. (This naturally becomes an ongoing process once a firm begins operating a warehouse network.) At the tactical level, a firm must concern itself with such planning activities as balancing the demand for warehousing capacity across its network, and planning the most efficient and effective utilization of its capacity at each individual DC. Capacity planning at the individual DC level can involve determining the overall labor level and mix required to meet the projected demands over the planning horizon, the proper mix and use of available storage locations (e.g., type of racking where adjustable), and so on. In general, tactical warehouse planning, similar to tactical production planning, focuses on the determination of how to best employ the existing network infrastructure (*i.e.*, the existing warehouses and material handling equipment). Additionally, decisions to purchase relatively minor additional warehousing assets (e.g., incremental material handling equipment, racking, etc…) will occur in the tactical planning process. Major infrastructure issues that a firm cannot resolve at the tactical planning level (e.g., inadequate network capacity to meet forecast long term warehouse throughput or storage requirements) must typically be fed back up to the strategic planning level for resolution. Thus, the efficacy of hierarchical warehouse planning and scheduling relies upon feedback loops, similar to the dependency of effective production planning on such mechanisms.

As Figures 5.1 and 5.2 illustrate, there exist a wide array of additional issues that a firm must address at the tactical warehouse planning level. In fact, Figure 5.2 provides just a small sampling of key questions that the tactical level must consider. Rather than repeat the list shown in Figure 5.2, we leave it to the reader to review this figure. An important point here remains that issues and questions that planners at the tactical level cannot completely address must be sent back to the strategic planning level. In the absence of this feedback approach, sub-optimal strategic decisions will occur and/or decisions that require resolution will linger because insufficient resources are available at the tactical planning level to address what are really strategic decisions.

At the operational level, a broad assortment of warehouse planning and scheduling activities take place on a regular basis. Figure 5.2 illustrates a sample of the decisions that operational schedulers must address. The scheduling of labor and short-term assignments of items to storage locations represent two of the major operational planning activities. Typically, it is the non-routine components of these activities (e.g., addressing temporary labor or storage requirements that significantly exceed capacity) that require the most critical attention. It is also typically the "exceptions" or "non-routine" requirements of operational planning and scheduling that planners must report or "feed back" to the tactical planning level. For example, when warehouse planners consistently find themselves having to schedule "unplanned" outside storage because of insufficient facility storage capacity, they should send this information to the tactical level for resolution. Perhaps the overall warehouse network is out of balance and requires re-alignment because excess storage capacity exists at certain warehouses, while other warehouses face the opposite situation. Alternatively, perhaps this storage capacity issue at one warehouse is not an imbalance issue, but rather is occurring regularly across the network and requires a total network solution. This represents just one simple example of the types of feedback loops that must exist between the operational and tactical warehouse planning levels.

Sections 5.2 and 5.3 will offer additional discussion of operational planning level activities. For now, we close this discussion by again emphasizing the importance of the feedback loops illustrated in Figure 5.1.

Objectives Of Hierarchical Planning Approaches For A Warehouse Network

A hierarchical planning and scheduling approach helps to assure that a firm's warehouse network carries out its mission and objectives. Figure 5.3 summarizes some of the traditional objectives that firms expect an effective and efficient warehouse network to meet. In addition to those shown in Figure 5.3, a firm generally will have several other objectives related to the specific mission of its network. The planning examples that we will now review in sections 5.2 and 5.3 are illustrative of the types of techniques that can help a firm achieve the objectives described in Figure 5.3.

5.2 Integrating Warehouse Planning With Production And Distribution Planning Activities

This section focuses on approaches for effectively linking warehouse planning activities and the warehouse network with the other key components and planning activities of a manufacturing and distribution network. We begin by presenting a framework to facilitate this desired integration. Next we review in additional detail a critical component of this framework, the DC (or warehouse) workload model. This review will illustrate how supply chain planning activities such as forecasting and inventory planning can be directly linked with warehouse planning through a DC workload model.

- **Facilitate high productivity rates in long-term and day-to-day material management activities**
- **Facilitate effective use of labor and equipment**
- **Maximize the utilization of warehouse space without creating congested working conditions**
- **Support efficient stock location and identification activities**
- **Provide for effective stock picking and order filling activities**
- **Promote the efficient unloading and loading of inbound and outbound shipping vehicles**
- **Facilitate efficient cycle counting and warehouse physical counts**
- **Promote accurate inventory record-keeping**

Figure 5.3 Illustrative Objectives For Warehouse Network Design And Planning Activities

Figure 5.4 presents a framework that integrates warehouse planning with the generic hierarchical production and distribution planning system illustrated in Chapter 1. It is important that a firm maintain linkages between its warehouse network and the other elements of the supply chain at the strategic, tactical and operational levels. Figure 5.4 illustrates an integration approach for the operational level in particular, and we will begin at this level and then address the tactical and strategic levels.

The operational forecasting, production and distribution planning system must provide regular updates to the warehouse planning and scheduling system. Critical data elements that the warehouse planners require include:

- Planned shipments from plants and other suppliers (inbound) to a DC, by week, by sku and product family[2],
- Planned shipments (outbound) from a DC to its customers, by week, by sku and product family, and
- Projected inventory at a DC, by week, by sku and product family

As Figure 5.4 depicts, the DRP/inventory management system represents the most likely planning tool to provide this data to the warehouse planning system. As reviewed previously, the operational forecasting system should provide forecasts by sku over the operational planning horizon to the DRP/inventory management system (*i.e.*, typically over a rolling 12 to 18 months). The DRP system evaluates the forecasts by sku, by DC, as well as the existing and projected inventory positions of the DC (and overall network). This evaluation facilitates the communication of projected net requirements to the production planning and scheduling system which in turn evaluates the ability of the plants to produce the requested DRP net requirements. Through an iterative process between the DRP

[2] Note that the appropriate time periods and product level (e.g., item or product family) will vary from one firm and/or industry to another.

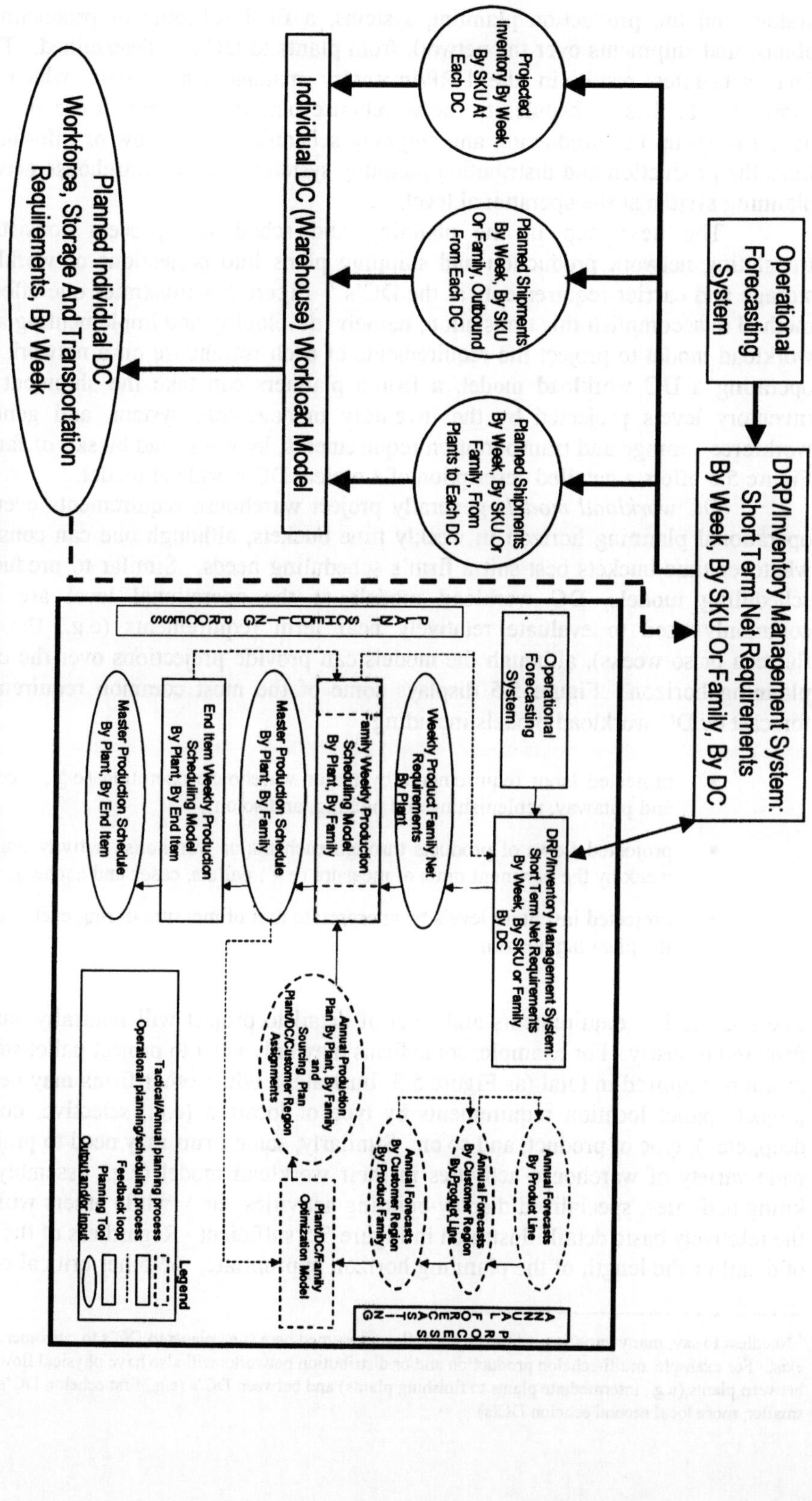

Figure 5.4 Integrating DC Workload Models With A Hierarchical Production And Distribution Planning System

system and the production planning systems, a final schedule of production at plants, and shipments over the network from plants to DC's is determined. These final schedules reside in the DRP/inventory management system which can communicate these schedules to the warehouse planning systems, once planners have firmed up the production and shipping schedules. This flow of information links the production and distribution planning systems with the warehouse network planning system at the operational level.

The next step in the planning and scheduling process consists of translating network production and shipping plans into projections of workload, storage and carrier requirements at the DC's.[3] Figure 5.4 illustrates one effective method to accomplish this translation; namely, developing and implementing a DC workload model to project the requirements of each warehouse on a network. By operating a DC workload model, a firm's planners can take the shipment and inventory levels projected by the inventory management system, and generate workforce, storage and transportation requirements, by week, and by sku or family. Figure 5.5 offers a detailed description of a typical DC workload model.

DC workload models generally project warehouse requirements over the operational planning horizon in weekly time buckets, although one can construct whatever time buckets best suit a firm's scheduling needs. Similar to production scheduling models, DC workload models at the operational level are most commonly used to evaluate relatively near term requirements (e.g., the next thirteen or so weeks), although the models can provide projections over the entire planning horizon. Figure 5.5 displays some of the most common requirements forecast by DC workload models including:

- projected labor requirements by major warehouse operation (e.g., receiving and putaway, replenishing and picking, and so on),
- projected flows of products through each major warehouse activity during a week by the pertinent units of measure (e.g., pallets, cases and eaches), and
- projected inventory levels by appropriate unit of measure during each week of the planning horizon

The specific DC requirements and level of detail to project will naturally vary by firm and industry. For example, some firms may only need to project pallet storage locations required in total (as Figure 5.5 illustrates), while other firms may need to project pallet location requirements by type of location (e.g., selective, double-deep, etc..), type of product, and so on. Similarly, some firms may need to project a wide variety of warehouse activities in their workload model (e.g., assembly and kiting activities, specialized display-building activities, etc.), while others will find the relatively basic detail illustrated in Figure 5.5 sufficient. Regardless of the level of detail or the length of the planning horizon appropriate, the other critical output

[3] Needless to say, many variations of the physical flows assumed here (*i.e.*, plants to DC's to customers) exist. For example, multi-echelon production and/or distribution networks will also have physical flows between plants (e.g., intermediate plants to finishing plants) and between DC's (e.g., first echelon DC's to smaller, more local second echelon DC's).

that a DC workload model must provide consists of a projection of the surplus or deficit in projected capacity, by time bucket, by activity or requirement. This represents the key output of a DC workload model and the information that will allow operational planners to run their warehouses efficiently. Additionally, as described in section 5.1, this also represents the critical information that warehouse planners need to feed back to the production and inventory planners to assure well-coordinated network-wide operations. For example, suppose the model for a particular warehouse indicates that the receiving and putaway functions of that warehouse will have a major capacity deficit that cannot be alleviated by temporary labor or overtime for weeks 5 through 9. This could result in a backlog of inbound trucks in the receiving yard, and more importantly, the unavailability of the needed level of inventory to fulfill customer orders at the warehouse during these weeks. In such a situation, the firm may have to divert planned inbound shipments to that particular DC, or in some other way address the potential problem. Without appropriate feedback from the warehouse planning function to the production and distribution planning functions, this potential disruption to network plans and operations may in fact occur. Thus, feedback *across* the various operational planning functions of the network represents yet another factor critical to assure efficient and effective overall operations.

DC workload models can also provide valuable tactical and even strategic planning decision support. Tactical and strategic DC workload models are quite similar to operational DC workload models in terms of content and methodology. However, planners typically are evaluating longer term questions such as the capability of an overall network (or a stand-alone warehouse) to meet projected capacity requirements over a planning horizon of two or more years, when they utilize tactical/strategic DC workload models. Interestingly, long run DC workload models often must still evaluate "daily" or "weekly" capacity requirements in order to produce insightful planning results. This is particularly true in the case of warehouse activities such as picking and shipping where volumes can differ dramatically by day of the week, week, month or season. In an environment where volumes swing dramatically, it becomes critical to accurately model the peaks and valleys of activities so that one can develop an accurate assessment of the appropriate capacity to maintain (*i.e.*, averages will not provide the necessary insights). Generally, in such an environment, it is not economical to establish the capacity levels of warehouse activities or storage areas such that capacity always meets or exceeds requirements. Rather, it is often more economical to plan to require temporary labor and/or storage during certain "peak" periods rather than to set staffing and/or storage levels high enough to accommodate the absolute peak requirements. A tactical or strategic DC workload model that does not evaluate these peaks correctly will fail to provide the proper decision support.

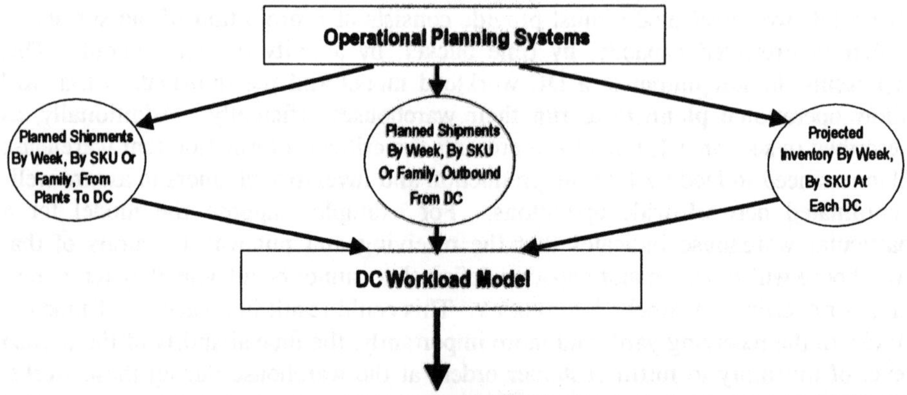

	Week 1			Week 2	. . .
	OPERATIONS				
	Receiving and Putaway	Replenishment and Picking	Packing and Shipping		
Projected Workforce Requirements					
Current/Planned Workforce Capacity					
Surplus (Deficit) Workforce Capacity					
Projected Throughput (Flows) - pallets - cases - eaches : .					
Current/Planned Throughput Capacity - pallets - cases - eaches :					
Surplus (Deficit) Throughput Capacity					

	STORAGE				
	Pallets	Cases	Other Units		
Projected Storage Requirements					
Current/Planned Capacity					
Surplus (Deficit) Capacity					

Figure 5.5 Illustrative DC Workload Model

Tactical and strategic DC workload models should fit into an overall integrated supply chain planning framework in a similar fashion to operational workload models. These higher level DC models require forecasts, inventory projections, and so on, just as do operational models. The framework will differ from the operational framework in that more aggregate, higher level forecasts (e.g., annual forecasts) and more aggregate inventory projections (e.g., annual or quarterly average or end of period levels) will typically drive tactical/strategic models. Therefore, part of the challenge in operating effective tactical DC workload models is often to translate aggregate level forecasts and projections into more granular forecasts (e.g., daily time periods) for model simulation purposes. Finally, it is also interesting to note that strategic DC workload models often are most effectively used in conjunction with a series of other network decision support models evaluating overall network requirements. Figure 5.6 displays one typical role of DC workload models in a strategic network planning study.

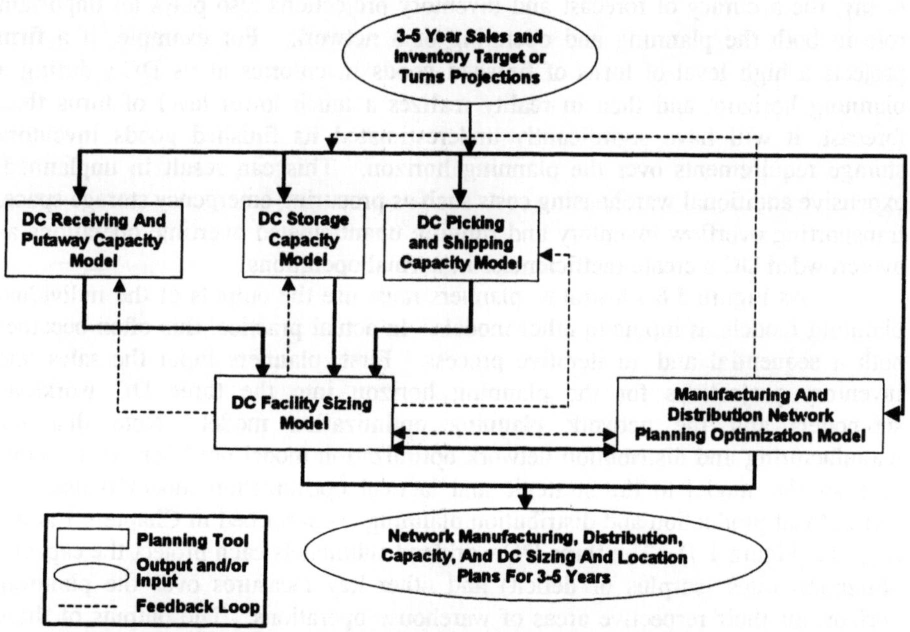

Figure 5.6 Illustrative DC Network Strategic Capacity Planning Analysis

The DC workload model in Figure 5.6 actually consists of three submodels shown individually in this diagram; namely, the receiving and putaway model, the storage model and the picking and shipping model.[4] These three submodels collectively serve to evaluate the capacity requirements and projections for the DC's on a network over a future planning horizon (three to five years in this example). As the figure depicts, in an ideal situation, this DC capacity analysis

[4] We note again that in actual practice, the appropriate number of submodels (or components) of a DC workload model will vary by firm, industry and particular application.

represents one element of an overall integrated methodology. For illustrative purposes, we now briefly review a typical strategic planning approach that utilizes the set of decision support models shown in Figure 5.6. In this example, we assume that a firm has a manufacturing and distribution network in place, and that it needs to evaluate whether its current warehouse network has sufficient capacity to address the projected demands for the next three to five years.

Illustrative DC Network Strategic Capacity Planning Analysis

A sales forecast for the strategic planning horizon naturally must drive a network capacity planning exercise. Often a "sales forecast" really will consist of a series of alternative forecast scenarios, each with a projected probability of occurring. Additionally, projections of inventory turns or target levels represent critical inputs to this planning process. The greater the level of inventory that a firm maintains on a DC network, the greater is the storage space that the network will require to store this inventory. Therefore, inventory target level projections, or alternatively, inventory turn projections are essential to an effective planning process. Needless to say, the accuracy of forecast and inventory projections also plays an important role in both the planning and operating of a network. For example, if a firm projects a high level of turns of finished goods inventories at its DC's during a planning horizon, and then in reality realizes a much lower level of turns than forecast, it will have significantly underestimated its finished goods inventory storage requirements over the planning horizon. This can result in unplanned, expensive additional warehousing costs such as procuring emergency storage space, transporting overflow inventory and running unanticipated overtime operations as overcrowded DC's create inefficiencies in normal operations.

As Figure 5.6 illustrates, planners must use the outputs of the individual planning models as inputs to other models. In actual practice, this often becomes both a sequential and an iterative process. First, planners input the sales and inventory projections for the planning horizon into the three DC workload submodels and the network planning optimization model.[5] Note that the manufacturing and distribution network optimization model used here is the same or a similar model to the strategic and tactical optimization model(s) used for hierarchical production and distribution planning, as described in Chapters 1 and 2 (e.g., see Figure 1.7). The three DC workload submodels each project the capacity utilization rates (surplus or deficit) and other key measures over the planning horizon for their respective areas of warehouse operations. The outputs of these three models become inputs to the DC facility "sizing" model. The DC facility sizing model evaluates such factors as the total network-wide warehouse square footage required to receive, store, pick and ship projected sales and inventory over the planning horizon. This model takes on particular importance if the results

[5] For purposes of simplicity, we simply refer here to sales and inventory forecasts being input to the network planning models. In fact, the level of detail of these forecasts is an important consideration that must be weighed carefully. For example, in certain applications, it may only be necessary to input forecasts at a major product line level (e.g., a sales division) for certain key years of the planning horizon (e.g., the first and last years). Alternatively, in other cases, more detailed product forecasts may be required for each year and a dynamic (*i.e.*, multi-period) network optimization model may be required. The level of granularity required must be evaluated on an application by application basis.

from the three DC workload models indicate a large deficit or surplus of warehouse capacity at certain points during the planning horizon.

The manufacturing and distribution network optimization model provides a manufacturing and distribution sourcing and supply plan for the planning horizon. This plan indicates forecast flows of product and materials through the network by individual location, from plants to customers. Thus, this model projects the future product and materials volumes that each warehouse will have to handle. In some applications, planners will need to run this model first before running the DC workload models so that the projected portion of total demand that each DC will have responsibility for is clearly identified. In other cases such as pictured in Figure 5.6, an existing supply and distribution plan may be in place, and planners can input the initial planning horizon forecast directly into the DC workload models. In either case, an iterative planning process must ensue whereby the implications of results from the DC workload and facility sizing models must be evaluated in the network optimization model, and vice-versa. (The feedback loops shown in Figure 5.6 illustrate these potential informational flows.) As noted, the iterations and feedbacks between models take on particular importance when there are either predicted significant capacity shortfalls or excesses during the planning horizon. For example, suppose that the initial results from the DC capacity and sizing models predict that the total warehouse network square footage will have to increase by 50% during the planning horizon. Further, assume that the firm can address this expansion requirement by either expanding one or more warehouses on its existing network, or by opening a new warehouse (*i.e.*, increasing the total number of DC's on the network). At this point, such factors as transportation costs to deliver products to customers, potential warehouse economies of scale and technology selection decisions that are affected by both the size and locations of individual warehouses come into play. Addressing a complex set of issues such as this necessitates the iterative, integrated modeling approach outlined in Figure 5.6.

In terms of methodology and technique, there exist a wide range of options to construct the types of models depicted in Figure 5.6. For illustrative purposes, we conclude this section by briefly considering one commonly used set of methodologies. Among other purposes, this discussion serves to make the point that with simply spreadsheet and database software, supplemented by a good optimization software package, planners can construct large scale integrated systems to perform complex network planning evaluations.

The DC workload models shown are typically built as either simulation models or sophisticated spreadsheet models that have some limited simulation capabilities.[6] Although the DC workload models may evaluate a long term planning horizon (e.g, several years), these models must be able to evaluate the peaks and ebbs of daily demands on a DC. Thus, as noted, these models must have at least limited simulation capabilities, and often a cleverly constructed

---

[6] By simulation model here, we do not necessarily refer to "dynamic" simulation models in the sense of traditional simulation models that have dynamic feedback loop mechanisms to evaluate a process (e.g., an order processing call center). Rather, we rather loosely refer to simulation models here as models that can evaluate a flow of daily receipts, shipments, etc. viewed at an aggregate level (e.g., by day) through a mathematical calculation.

spreadsheet, supported with an appropriate database can facilitate the development of a solid DC workload model. Naturally, the option always exists to develop more sophisticated simulation models should the planning environment require this. The manufacturing and distribution network model almost always requires the use of optimization and/or optimization heuristics, as illustrated in the first three chapters. Finally, the DC facility sizing model can often be constructed as a sophisticated spreadsheet supported again by a good database. This model must accurately quantify the facility space (e.g., square footage) required to support an overall network, as well as individual warehouses on a network. Typically this requires a well-defined set of linear and non-linear algebraic equations that can perform such tasks as translating projected storage requirements described in units (e.g., cases or pallets) into measures such as square feet of warehouse space required to inventory the projected units. Here engineering based data and equations on such factors as storage racking space requirements and so on must be modeled so that one can develop the desired space projections in this example. Similarly, these models should have equations to quantity the costs of either expanding or contracting the size of the existing network in total, and/or individual locations on the network. These cost equations should have the capability to differentiate for such factors as the relative cost of operations of alternatively sized warehouses (*i.e.*, economies of scale). Without belaboring this discussion further, one can see that a well-constructed spreadsheet can in fact provide sufficient support to develop a good DC Facility Sizing model. As noted, there exist numerous techniques for developing the set of models described in Figure 5.6. However, the aforementioned approach certainly represents a very adequate modeling approach for DC network strategic capacity planning.

5.3 A Hierarchical Warehouse Facility Design And Operations Planning Example

Section 5.3 concludes this chapter with a review of an actual implementation of a hierarchical warehouse facility design and operations planning example. The authors who developed this approach (Gray, Karmarkar and Seidmann, 1992) employed this methodology to assist an automotive manufacturer and distributor in designing and operating a medium-sized warehouse (100,000 square feet). This application holds particular interest because, as described by its authors, it relied heavily on an iterative, hierarchical solution methodology to select the ultimate design, operating policies, etc.[7]

This case study addressed decisions ranging from warehouse layout and design, to materials handling and storage equipment choices, to the determination of operating procedures intended to guide daily warehouse operations. In approaching this case study, the authors described the problem that they faced, as well as their rationale for employing a hierarchical planning methodology, as follows:

---

[7] This section is based on the description of this implementation provided in Gray, Karmarkar and Seidmann, 1992.

> We consider(ed) the broad problem of warehouse design and operation, including warehouse layout, equipment and technology selection, item location, zoning, picker assignment, pick-list generation and order batching. Solving the entire problem formally with a single monolithic model appear(ed) to be impractical even in a small case. This consideration led us to formulate a comprehensive hierarchical decision structure for the problem of designing and operating order consolidation warehouses. Our hierarchical structure identifies and integrates the functional relationships among the different decision levels, with the decision variables grouped into a nested or hierarchical order. This... approach is based on a series of coordinated mathematical models which interlink the tradeoffs involved... (and which) serve to identify and evaluate superior design alternatives.

Thus, in explaining the need for a hierarchical approach to this warehouse problem, the authors' cite similar factors to those previously used to describe the rationale for employing hierarchical production and distribution planning systems. Specifically, the full set of issues and decisions to address represented such a complex problem that the authors had to decompose the problem into a series of more manageable (and mathematically tractable) linked sub-problems. In terms of the framework presented in Figure 5.1, this case study encompasses all three decision-making levels shown (*i.e.*, the strategic level through the operational level). However, this case study begins at the individual warehouse level, and therefore does not include the initial component shown at the strategic level, namely, the network design and location problem. Thus, this case study focuses on the design and operations of a single facility only. Finally, note that in defining the operating procedures to be addressed, the authors included such issues as "the implementation of zoning as well as picking schemes such as batch picking, simultaneous sorting, order consolidation and picking to different material handling media (such as carts and conveyors)". Therefore, in terms of operating procedures, this case study addressed what could also be termed the operating policies that run a warehouse. This case study does not address such issues as the ongoing daily scheduling of warehouse activities (*i.e.*, decisions at the lowest level of planning in the overall hierarchical scheme).

Gray et. al. (1992) detail the overall objective and scope of their case study through a qualitative description of the objectives, problem parameters, decision variables and problem constraints included in their application. Figure 5.7 displays this description.

Objective

Minimize the annualized incremental initial costs plus the warehouse operating costs. The incremental initial costs include:

(i) Warehouse layout rearrangements.
(ii) New storage hardware purchasing and installation.
(iii) Additional material handling systems purchasing and installation.
(iv) Defining, procuring and installing new information systems (data collection, software and computer hardware)

The operating costs include:

(i) Labor costs for:
- picking;
- sorting;
- replenishing;
- checking, auditing, and invoicing;
- picklist generation and exception handling.

(ii) Inventory holding costs.

Problem parameters

(i) Number of line items (SKU's).
(ii) Expected number of orders per day.
(iii) Demand per item per order.
(iv) Item dimensions.
(v) Labor costs ($/person/day).
(vi) Available storage facilities.
(vii) Current storage allocation policy.
(viii) Replenishment costs per storage technology.
(ix) Picking costs.
(x) The required mean length of a working day.
(xi) Picker pick times.
(xii) Picker walk times.
(xiii) Picker grab times.
(xiv) Sorting times.
(xv) Packaging and invoicing times.

Figure 5.7 Description Of Warehouse Facility Design And Operations Planning Case Study

Decision variables

The decision variables (grouped into levels) are as follows:

(i) Warehouse floor plan: Warehouse dimensions and the physical layout of aisles, racks and shelves. This determines overall material flow in replenishment as well as picking.

(ii) Equipment and technology selection: The choice of specific types of racks and shelving, storage equipment including automated storage, material handling vehicles, conveyor types and layouts, bar coding and reading equipment, and sorting equipment.

(iii) Allocation of items to racks and storage types.

(iv) Determination of item facings and total length of racks and shelves of each type. 'Facing' is the width of storage assigned to an item, typically expressed as multiples of a physical dimension of the item.

(v) Assignment of items and storage types to zones.

(vi) Number of pickers and number of zones.

(vii) Number of orders batched per pick cycle.

(viii) Number of sorters or sorting equipment.

(ix) Pick list generation method and sorting procedure.

Problem constraints

(i) All items must be assigned to storage facilities.

(ii) Each item must be assigned to a) at least one storage technology, and b) a unique position within one zone. Each storage position can hold at most one item.

(iii) Expected pick time per zone per day must be less than or equal to a given limit

(iv) The number of pickers must be greater than the number of zones. (Each zone is assigned at least one picker.)

(v) The assignment variables, the number of pickers, and the number of zones must be integers.

(vi) Sorting effort = f_1 (picking method, batch size, material handling hardware, sortation hardware, product identification scheme).

(vii) Pick time = f_2 (number of zones, number of pickers per zone, picking method, batch size assignment of items to storage technology, number of facings per item).

(viii) Replenishment effort = f_3 (assignment of items to storage type, number of facings per item)

(ix) Picking cycle time = f_4 (synchronization method used, number of zones, batch size, imbalance effect).

(x) Storage hardware cost = f_5 (assignment of items to storage technology, required number of facings per item, currently available storage hardware, the marginal cost per unit of additional storage capacity per storage technology).

(xi) Aisle length = f_6 (number of zones, assignment of items to storage type, number of facings per item, allocation of items among zones).

(xii) Aisle length $\leq$ Building constraints.

Source: Gray et. al., 1992

Figure 5.7 Description Of Warehouse Facility Design And Operations Planning Case Study (continued)

As noted, the authors address this multi-faceted problem through a hierarchical decomposition approach. In particular, Gray et. al. separate the problem described in Figure 5.7 into the following three levels:

1. Facility design and technology selection,
2. Item allocation, and
3. Operating policy

In the authors' view, the principal decisions to be made at each level for their particular application included:

- *[Level 1]* Selecting the major automation equipment to utilize by trading off the (amortized) initial purchase price of equipment with the annual operating costs associated with different levels and types of automation,
- *[Level 2]* Determining how to deploy items in aisles and zones, and
- *[Level 3]* Determining two key operating parameters, namely, the most efficient order batch size and the optimal number of warehouse zones.

The solution methodology employed for this problem consisted of an iterative process that spanned all three levels and eventually converged upon a solution that minimized total cost.[8] Gray et. al. summarized this process and the key data requirements necessary to support it as follows:

> The major exogenous data requirements are demand data, item attributes (volume and package size), picking and sorting times, and cost information. The Facility Design & Technology Selection decisions are made first and impose constraints within which more detailed decisions about Item Allocation and Operating Policy are made. These detailed decisions, in turn, provide the necessary feedback for evaluating the merits of the higher level decisions. We develop an iterative procedure of using a series of coordinated analytical models as a first cut at determining design and operating parameters. The result of the analysis is then taken as a base scenario, which directs further search to a small neighborhood of superior alternatives. The final search is carried out by using a detailed simulation, which is able to account for the many secondary issues which are simplified away in the analytical process. The use of analytical models drastically reduces the number of alternatives to be simulated; this makes the use of detailed simulation a viable procedure for fine-tuning the final solution.

[8] The reader interested in the mathematical details of the methodology is referred to Gray, Karmarkar and Seidmann (1992).

There are several interesting points regarding hierarchical planning that the authors touch upon here. First, their overall approach, similar to other true hierarchical systems, employs a top-down and bottom-up approach. Specifically, they allow decisions (or in this case modeling results) that occur at higher levels to impose constraints on decisions and solutions at lower planning levels. They similarly rely on decisions and solutions at the lower levels to provide feedback to higher levels. This feedback provides additional insights as to whether the original decisions made at the higher level were good, or whether some modification to the original decision is required. The authors also note that they perform *additional* analyses (*i.e.*, "detailed simulations) at certain levels which are above and beyond that performed in the initial simplified hierarchical planning model. This serves to illustrate that in a hierarchical planning system, there are typically some planning and scheduling activities that occur which are not explicitly linked to lower and/or higher levels. Rather, the decisions, schedules and so on that are explicitly linked across hierarchical planning levels essentially serve to provide a boundary or framework within which other more detailed decisions at a particular level occur without explicitly imposing constraints downward and/or feedback data upward.

Figure 5.8 depicts the actual hierarchical system developed by Gray et. al. to design a spare parts warehouse for an automotive manufacturer and distributor, and to establish the operating policies to guide the facility's operations. In this particular application, the authors' were redesigning an existing warehouse; and therefore, as Figures 5.7 and 5.8 show, their cost objective consisted of minimizing "incremental" facility redesign, new equipment and information systems related purchase costs, as well as ongoing annual operating costs. Clearly however, this methodology would also apply to a new facility and the objective function costs would simply change from "incremental" one-time costs (such as facility and new equipment purchase costs) to the full one-time costs to open a new facility.

The numerous individual analyses described in Figure 5.8 such as "assign items to zones (AIZ)" in planning level 2 and determine the "number of pickers and zones (NPZ)" in level 3 collectively address all of the components of the descriptive problem outlined in Figure 5.7. For illustrative purposes, Figure 5.9 provides an overview of the methodology used by Gray et. al. to make the operating policy decision on the number of pickers and zones to utilize in the warehouse. The actual model used at this planning level consisted of a series of straightforward algebraic equations that expressed certain key factors that would determine the number of pickers and zones required such as walk distance, total available time per day per warehouse employee, unload and obtain new pick list time and so on. (The reader interested in these explicit details is referred to Gray et. al., 1992). A quick review of this decision at the third and lowest planning level in the process illustrates the hierarchical paradigm employed here. In particular, before a planner can run this sub-component of the decision-making methodology, a number of decisions or scenario planning assumptions must have occurred.

At the initial planning level, decisions must have taken place regarding the physical layout of the warehouse, the aisle design and picking path routes, the equipment and storage technology that will be employed and so on. Similarly, at the next level decisions regarding aisle length, some preliminary assignment of items to zones, some preliminary assumption regarding the number of zones, and so on must take place. Then at the third level of the hierarchy, the authors' sequential approach required a determination of the order batch size. With all of these decisions made, or more accurately, decisions and/or assumptions made for the particular planning scenario; then the planners could run the analytic model that would determine the number of pickers required, and the number of pickers per zone required. As Figure 5.8 highlights, results of the analytic simulation to determine the number of pickers and additional new information on the number of zones required would flow back to the higher planning level. This would facilitate another planning scenario simulation that would again employ a sequential set of analytic models to determine a potential planning solution (*i.e.*, the design of the warehouse and its related operating policies). The cost equations in the series of analytic models would capture the total costs, both one-time and annual, associated with each planning scenario. As described by the authors, this process would eventually converge in a solution that identified a low cost and/or efficient and effective plan to both design and operate the warehouse.

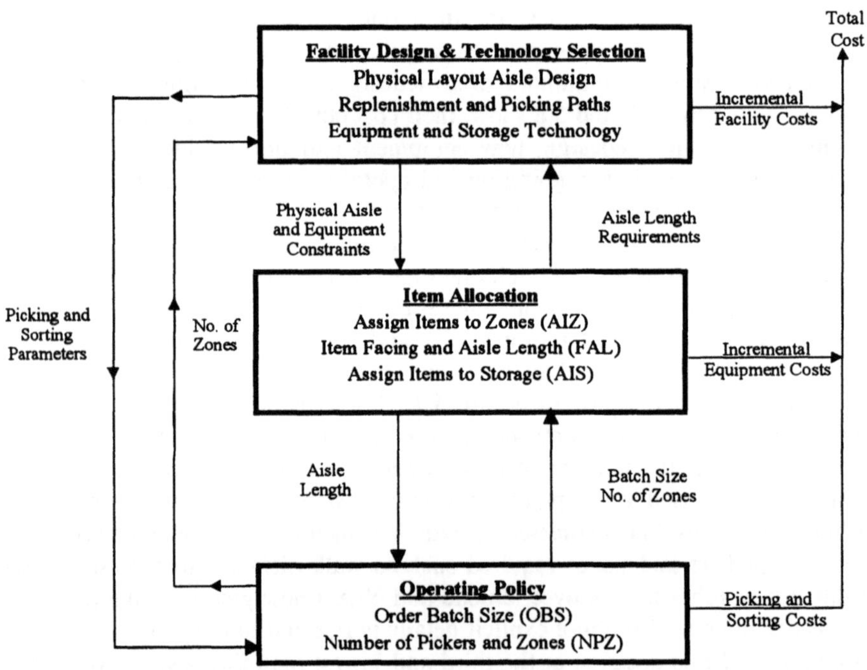

Figure 5.8 Hierarchical Analysis Of Warehouse Design And Operating Policies

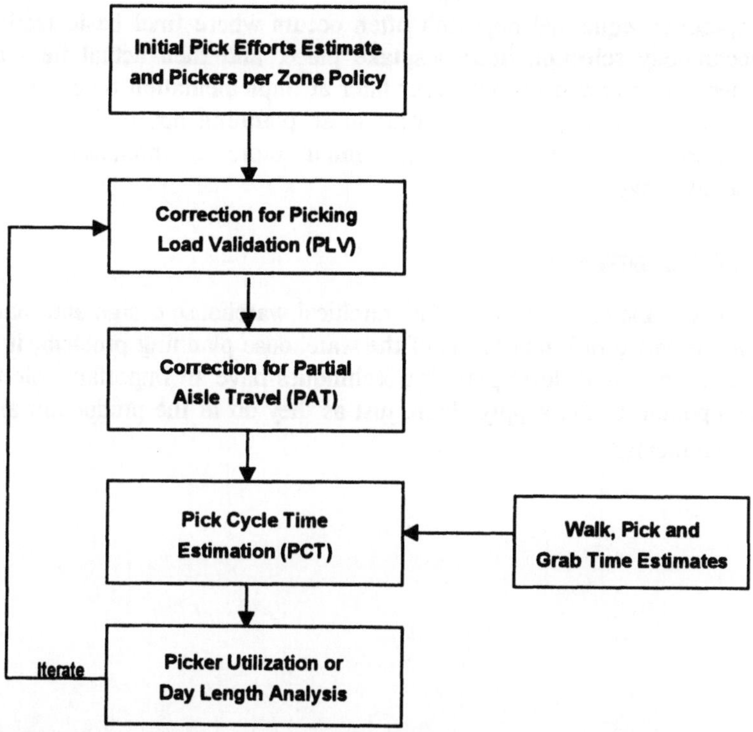

Source: Gray et. al., 1992

Figure 5.9 Determining The Number Of Pickers And Zones

We will not attempt to review each analytic submodel of this hierarchical warehouse planning system. However, this brief summary has served to highlight the utility of employing a hierarchical approach to the warehouse facility design and operations problem. In this application, by evaluating the potential impacts of higher level decisions on lower level decisions and operations, and feeding these potential impacts back up the planning process in an iterative methodology, the designers of this system uncovered numerous opportunities to reduce costs and improve operating efficiencies without damaging customer service levels. For example, Gray et. al. (1992) described that their:

> analysis of storage requirements result(ed) in establishing a total aisle length requirement that (was) 25% less than the (existing) layout. Furthermore, the study revealed that substantial reductions (on the order of 50%) could be made in picking, sorting and checking labor requirements, without loss of handling capacity, by using the improved layout in conjunction with the (new) zoning and batch policy parameters.

In closing this case study, we note that firms often make warehouse facility design decisions based on less detailed hierarchical analyses. In practice, a more top-down sequential approach often occurs where final basic facility design and technology selection decisions take place, and then actual item allocation and operating policy decisions occur later at implementation time. Nevertheless, it is important to keep this ideal warehouse planning approach in mind, and where appropriate to use this or a similar more comprehensive decision-making methodology.

5.4 Summary

This completes our review of hierarchical warehouse design and management. As we discussed different facets of the warehouse planning problem, it was clear that integrated, multi-level planning techniques have an important role to play in this component of the supply chain, just as they do in the production and distribution components.

Chapter 6

Inventory Management And Forecasting

Inventory management and forecasting are planning activities that go hand-in-hand. Both play critical roles in the management of a firm's supply chain and in hierarchical production and distribution planning systems. Further, the accuracy of a firm's forecasting process directly impacts the level of inventory that a firm must maintain to meet customer service order and line fill rate objectives. In this chapter, we examine selected "hierarchical" dimensions of the processes of inventory management and forecasting.

Section 6.1 begins with a review of how inventory decisions fit into an overall planning hierarchy and what types of decisions typically occur at each planning level. In section 6.2, we explore several traditional inventory management analyses and decision support tools often used to segment a firm's total finished goods inventory into smaller subgroups. This will include a review of traditional ABC inventory analysis techniques. Section 6.3 presents a simple hierarchical algorithm that facilitates effective annual and tactical inventory planning at the product family level, while maintaining visibility to end item and sku level related considerations. In section 6.4, we explore approaches for developing tactical and operational planning forecasts that utilize multiple levels of forecasts in order to assure an overall integrated product line forecast (e.g., top-down and bottom-up forecasting techniques). Section 6.5 examines how hierarchical forecasting and inventory management plays an intricate role in such activities as the Sales and Operations Planning process that have become so popular in recent years at manufacturing and distribution firms. Finally, in section 6.6, we offer some brief concluding comments.

6.1 How Inventory Decisions Fit Into A Planning Hierarchy

Similar to production and distribution supply chain decisions, inventory decisions encompass all three levels of the planning hierarchy (*i.e.*, the strategic, tactical and operational levels). Additionally, in a similar fashion to the production and distribution planning functions, inventory decisions made at higher levels impose constraints upon decisions that occur at lower levels; while at the same time, the ramifications of constraints imposed by higher level decisions must flow back or upwards to higher planning levels from lower planning levels. Figure 6.1 offers an overview of where some key inventory decisions fit into a planning hierarchy. While this set of decisions does not represent a comprehensive list, it serves to illustrate how key inventory decisions span multiple planning horizons. For illustrative purposes, we now consider several of these decisions in more detail.

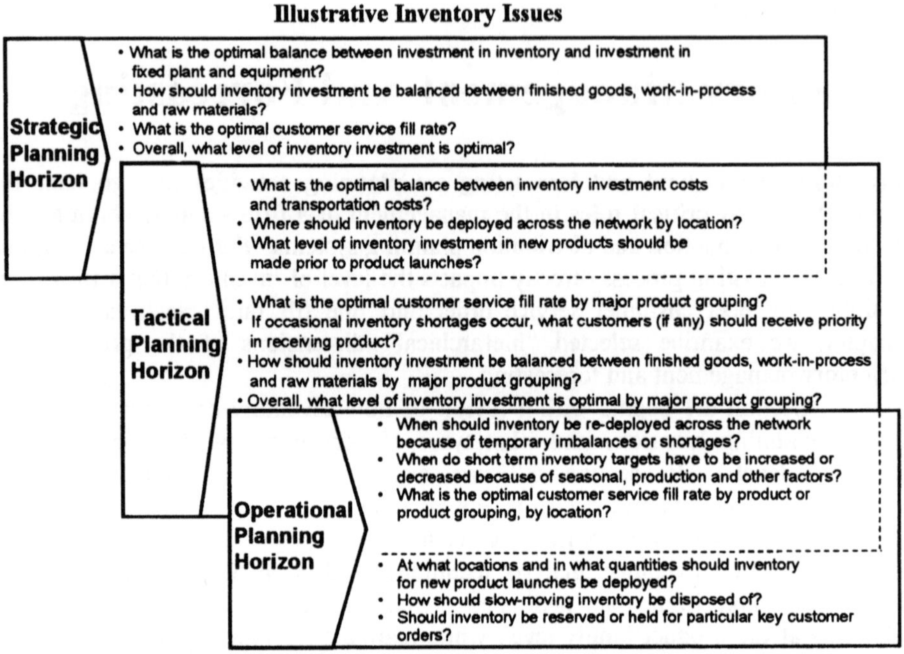

Figure 6.1 How Network Inventory Decisions Fit Into A Planning Hierarchy

At the strategic planning level, inventory decisions revolve around such questions as what is the optimal level of inventory investment that a firm should plan to maintain on an ongoing basis. This decision requires an evaluation of a number of trade-offs. For example, in many cases, manufacturing firms must consider what represents the best mix of investment in manufacturing capacity vis-à-vis total investment in inventory. As a general rule, for any given level of network-wide demand, as a firm increases its manufacturing capacity and flexibility, it decreases the level of inventory investment that it must make. This is particularly the case for make-to-stock firms. For example, if a firm has the capacity and flexibility to produce its entire product line during one week (*i.e.*, to cycle through its full set of products), it will require less finished goods inventory to fill customer orders than if the firm requires several weeks or months to cycle through its product line. In general, the more heavily a firm chooses to invest in fixed plant and equipment, the greater its production capacity and flexibility. Thus, at the strategic level, a manufacturer must decide the best balance to maintain between fixed assets (plant and equipment) and inventory investment.

The inventory investment decision also requires that a firm consider the trade-offs between relative levels of investment in finished goods inventory versus raw materials and work-in-process components. Additionally, at a very high level, a firm must develop a strategy on its planned customer service level fill rate. For example, a make-to-stock manufacturer cannot simply plan to have a 100% customer line fill rate. This is not realistic. Thus, a manufacturer must put a service level strategy in place considering the trade-offs between some "acceptable" level of lost or unfulfilled demands versus the inventory investment associated with alternative service levels. Finally, such issues as the appropriate trade-off between transportation costs and inventory investment costs also require consideration in the inventory management process. Often manufacturers must choose between using faster, more expensive transport services (e.g., air) between a "supplying" origin and a "receiving" destination and slower, less expensive transportation alternatives. A key question in such decisions is do the savings in inventory investment requirements (*i.e.*, in-transit and safety stock inventory at the destination) facilitated by a faster replenishment transit time outweigh the increased costs of the expedited transport service. Questions such as this often represent tactical rather than strategic issues, however, they play a prominent role in a firm's overall inventory investment approach.

At the tactical planning level, many of the same inventory issues found at the strategic level re-surface, but at a greater level of granularity. For example, a firm's annual planning process generally includes an evaluation of the total inventory investment that it will make during the next year. At the minimum, however, this process will usually include an analysis at the major product grouping level (e.g., brand or product family). (In contrast the strategic inventory planning process will not address this level of detail.) Many of the same trade-offs considered at the strategic level again re-appear in more detail (e.g., evaluating optimal service level fill rates and the appropriate mix of finished, in-process and raw materials inventory at the product grouping level). Further, as previously noted, decisions implemented at the strategic level will impact options at the tactical level. For example, the manufacturing capacity and flexibility built into the firm's current infrastructure will heavily influence its annual inventory investment plan.

At the tactical level other more detailed decisions such as the deployment of inventory by location and by product grouping will also occur for the first time. An example of the type of feedback that can develop at this level would be if a firm's planners, when reviewing inventory requirements by major product grouping, determine that in order to meet planned service fill rates, they require a total inventory level exceeding the planned overall investment target. Additionally, at this level, specific policies must be developed covering such potential issues as whether certain customers will receive priority in the event of temporary inventory shortages during the planning horizon. The development of guidelines for when inventory should be re-deployed between locations on a firm's network because of shortages and/or imbalances represents another example of the types of policies developed at the tactical planning level.

At the operational planning and execution level, again many previous planning decisions made at higher levels are revisited at a more detailed product line level and in more detailed time increments. Here a firm's planners must ensure that all sku's (*i.e.*, items at unique locations) have inventory targets designed to deliver specific customer service fill rates. Additionally, decisions regarding inventory targets by season, by month (and perhaps even by week) must be made, and must fit within the guidelines or constraints established at higher planning levels. Inventory policies represent a vital area where the existence of good feedback loops from the operational level to the tactical level play an important role. If for example the guidelines regarding inventory re-deployment between network locations in cases of inventory imbalances are not working effectively, this information must be communicated back to the tactical planning level so that more effective policies and procedures can be developed and implemented.

In summary, the inventory management function shares the same critical need for coordination and feedback between hierarchical planning levels as do the manufacturing and distribution functions. Further, effective inventory management represents an integral and prerequisite component of effective hierarchical production planning.

6.2 A Hierarchical View Of A Firm's Inventory

Perhaps the most traditional approach used in inventory management consists of identifying and managing inventory based on an ABC classification system. The concept of ABC actually dates back to the Italian economist Wilfredo Pareto and his studies of the distribution of wealth in Italy. Among other things, Pareto observed that about 20% of the population controlled about 80% of the wealth, and over time, this observation has led to the 80/20 rule used daily in an innumerable number of disciplines and practical methodologies. As noted, in the world of inventory management, the 80/20 rule has manifested itself in the form of ABC classifications commonly used by firms to stratify their inventory. Formally, an ABC system represents a ranking of the finished goods end items or sku's in inventory based on one or more measurement criteria. Criteria for establishing ABC rankings abound, however, the most common measures include:

- sales value (*i.e.*, total dollar sales)
- total unit sales
- total inventory investment value (at cost)
- frequency of sales, purchase or shipment
- total profit contribution, and
- total inventory storage space (e.g., cube) or pallet locations occupied

There exist numerous measures used to rank inventory because the appropriate measure may vary depending upon the particular application. For example, while total sales dollar value probably represents the most common measure used, a warehouse planner will also want to hierarchically rank inventory based upon the total storage and frequency of replenishment requirements placed upon distribution centers by individual items. Additionally, firms often apply ABC methods to many different dimensions or subcomponents of the supply chain. Planners often develop rankings for a firm's total network, for geographically-based sub-entities (e.g., countries or regions in the case of a large global firm), for major manufacturing and/or product line divisions, for individual inventory storage locations (e.g., warehouses), for brands within product line divisions, and so on. Further, firms employ ABC methods to help manage raw materials and work-in-process goods, as well as finished goods inventory.

Figure 6.2 presents a typical ABC example of the relationship between the number of end items in a firm's inventory and the relative contribution of end items to overall firm sales. Naturally in practice, the percentage of the total represented by A, B and C items will vary. However, one often observes that the top 20% of the items (*i.e.*, the A's) account for about 80% of total sales, the next 30% of the items (*i.e.*, the B's) provide about 15% of total sales, and the remaining 50% of the items (*i.e.*, the C's) represent about 5% of total sales. It is also a common practice to stratify inventory items into one or two additional classifications such as "D" items to represent particularly "slow-movers" or perhaps even items that literally have had no sales for some defined recent period of time (e.g., the last 12 months).

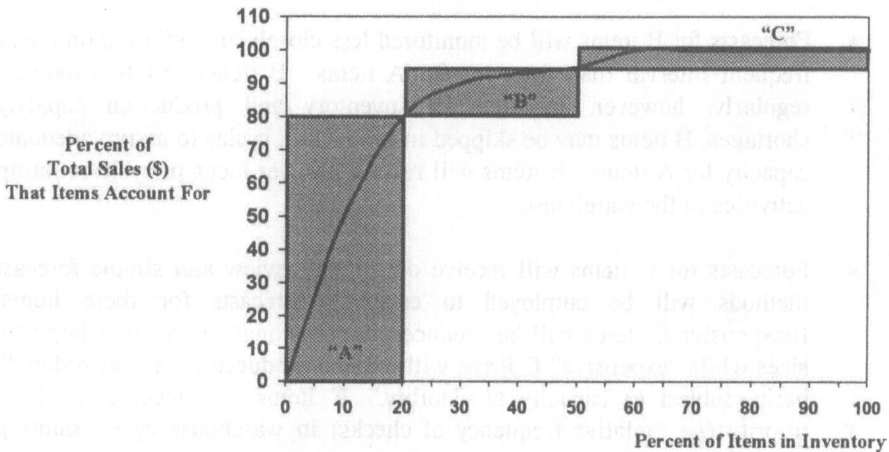

Figure 6.2 ABC Classification Of Finished Goods Items

A hierarchical classification of inventory can play a number of important roles in a firm's inventory management practices. First of all, it offers a valuable perspective on the relative utility that individual inventory items provide. This is particularly the case for the occasional firm that implements an ABC approach for the first time (*i.e.*, a firm not using ABC as a standard business practice) or a firm that implements a new application of ABC (e.g., a storage space analysis at an individual distribution center). Additionally, ABC stratifications often provide a logical basis for firms to develop and implement policies. Traditional examples include:

- establishing separate group service goals (e.g., line item fill rates) for A, B and C items based upon their relative importance (e.g., all A's 99%, all B's 98%, all C's 95%),
- establishing production priorities for items in cases where temporarily demand outstrips production capacity and a firm must allocate capacity to only certain items, and

- determining the level of resources devoted to items in such activities as forecasting, inventory cycle counting and production planning.

The following represents a typical set of policies for A, B and C items that a firm might establish:

- Forecasts for A items will be monitored closely and adjusted as appropriate on a regular basis. A combination of sophisticated forecasting techniques, augmented by current marketing and field sales input and top down vs. bottom up forecast checks will be used. A items will be produced regularly and receive production priority in times of inventory and production capacity shortages. Additionally, A items will receive priority in warehouse cycle counting activities.

- Forecasts for B items will be monitored less closely and adjusted on a less frequent interval than forecasts for A items. B items will be produced regularly, however, in times of inventory and production capacity shortages, B items may be skipped in production cycles to assure adequate capacity for A items. B items will receive a lower focus in cycle counting activities in the warehouse.

- Forecasts for C items will receive occasional review and simple forecast methods will be employed to establish forecasts for these items. Inexpensive C items will be produced in occasional, (relatively) large lot sizes while "expensive" C items will only be produced on an "as ordered" basis, subject to capacity availability. C items will receive the least priority (*i.e.*, relative frequency of checks) in warehouse cycle counting procedures.

The actual policies implemented based on ABC analyses will naturally vary by firm and industry. Nevertheless, the preceding fictitious policy set illustrates the types of operating guidelines that ABC analyses help firms and planners implement.

ABC based inventory management represents a different "hierarchical" dimension than traditionally thought of in hierarchical production and distribution planning. In this case, the hierarchy relates more to the relative importance of end items or non-finished goods inventory rather than to the dimension of "planning horizon" thought of in conventional strategic, tactical and operational hierarchical planning. However, the ABC hierarchical dimension represents an important view to consider in operations and supply chain management. Before we review another hierarchical planning dimension of inventory management in section 6.3, we conclude this section this with a brief illustration of another classification scheme for inventory management decision support.

In practice, one frequently finds that many firms have widely varying weeks of supply of inventory by item across their network. Sometimes this occurs by design; however, often this situation evolves over time for numerous reasons such as poor tracking of inventory by planners, unsuccessful promotions and unexpectedly short product life-cycles. Regardless, a regular review of the weeks of inventory in a supply chain can provide important monitoring information for a firm's inventory planners. Figure 6.3 illustrates an inventory classification scheme based upon weeks of supply in inventory. Note this figure displays the number of items within established weeks of supply intervals (e.g., between 5 and 10 weeks), as well as the total inventory investment represented by the items in an interval. This figure portrays one more example of the numerous "stratification" schemes commonly used by practitioners to monitor inventory.

6.3 Aggregating End Items For Family Based Planning

In this section, we consider another important element of inventory and production planning; namely, hierarchical analyses of inventory levels. A key step required to facilitate effective tactical (annual) and operational product family planning and scheduling consists of properly aggregating end item inventory levels and requirements into product family level planning methodologies. This step applies at both the network-wide tactical production planning level and at the individual plant tactical and operational planning levels discussed in Chapters 2 and 3. This section focuses upon one critical element necessary to assure the proper hierarchical link between end items and product families in production planning; namely, utilizing a good method for determining beginning-of-period inventory levels in annual (tactical) planning models. Specifically, we present a simple algorithm which evaluates and incorporates the current inventory positions of individual end items into long run, product family based production and distribution planning models. This algorithm lends itself to any product family based production and distribution planning technique (e.g., optimization, simulation, spreadsheet models, etc.). Further, one can apply this to any production or distribution planning technique that models products at a level of aggregation higher than the end item (e.g., methods for planning at the product type level).

Product Line Structure

Figure 6.4 illustrates the levels of product aggregation in a firm's planning process that will be assumed for the balance of this section. One can observe that end items tree up into product families, and product families then tree up into product lines. As noted, the algorithm presented here assumes that the firm's annual integrated production and distribution planning process models products at the product family level (level 3 in Figure 6.4). As discussed in Chapters 1 and 2, product families are created based upon the similarity of one or more key commonly shared characteristics of each end item within a family (e.g., the end items have similar production rates and costs, and are produced on the same lines).

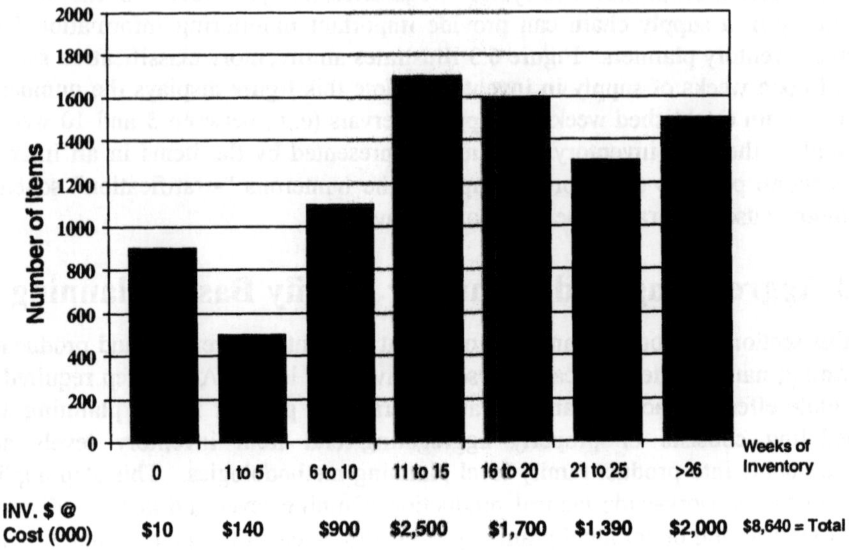

Figure 6.3 Illustrative Weeks Of Supply Stratification Scheme

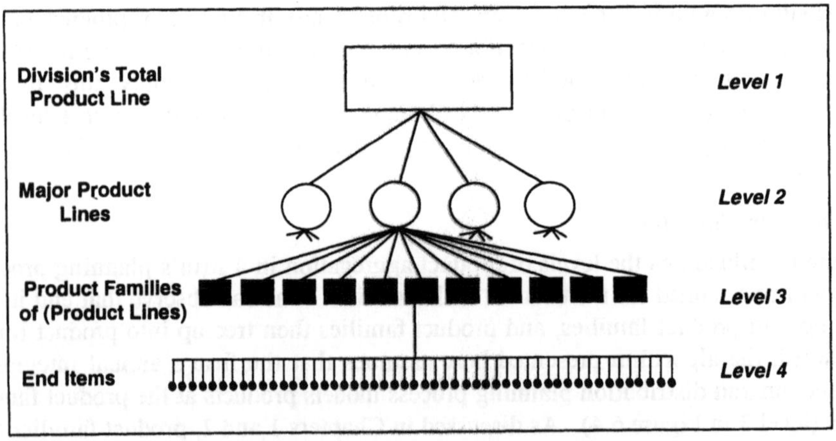

Figure 6.4 Illustrative Product Aggregation Hierarchy

Why Is It Critical To Incorporate Current End Item Inventory Positions Into Annual Product Family Level Production Plans?

It is necessary to evaluate current end item inventory positions over the entire manufacturing/distribution network in order to assure that the firm's annual production plans do not underestimate total manufacturing capacity requirements. Product family inventory data, when developed at the product family level (rather than at the end item level), can provide misleading information leading to this underestimation of capacity requirements in annual or any other planning models.

To clarify this concept, we will now consider the projected production requirements over an annual planning horizon for an illustrative product family when its beginning of period inventory is evaluated at two different levels of product aggregation: the end item level and the product family level. Before proceeding with this example, however, we introduce two definitions that serve to measure the utility of inventory over a planning horizon.

<u>Definition 1.</u> **Useable Inventory Over The Planning Horizon is defined as the minimum of:**

1. *(forecast annual sales over the planning horizon) + (end-of-period inventory objective or target), or*
2. *(inventory at the beginning of the planning horizon)*

<u>Definition 2.</u> ,**Excess Inventory Over The Planning Horizon is defined as the maximum of:**

1. *(the current or beginning of period inventory) - (useable inventory over the planning horizon), or*
2. *0.*

Figure 6.5 provides an illustration of these two definitions ior one end item. There are 13,000 units of inventory of this particular end item at the beginning of the planning horizon, and the end item has forecast sales over the planning horizon of 10,000 units and an end of horizon inventory target of 1,000 units. According to definition 1 then, there are 11,000 units of "useable" inventory over the planning horizon of this end item (*i.e.*, the minimum of: [10,000 + 1,000] and 13,000 is 11,000). Definition 2 thus would indicate that there exist 2,000 units of excess inventory over the planning horizon of this end item (*i.e.*, the maximum of: [13,000 - 11,000] and 0 is 2,000). With useable and excess inventory over the planning horizon now defined, we can proceed to evaluate the production requirements of an illustrative product family by two alternative approaches.

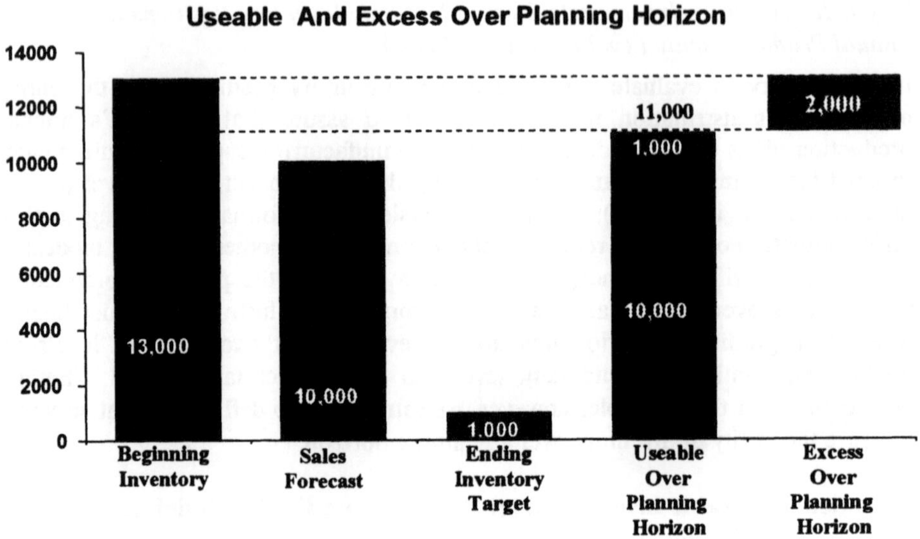

Figure 6.5 Illustration Of Inventory Definitions For One End Item

Table 6.1 displays an analysis of the production requirements for a product family calculated exclusively at the product family level. Note that this product family has 9,200 units in inventory at the beginning of the planning horizon, it has forecast sales of 16,000 units and an end of horizon inventory target of 3,200 units. Table 6.1 shows that with some rudimentary calculations we can determine that this product family requires 10,000 units of production over the planning horizon, and that all 9,200 units of beginning inventory are "useable".

Table 6.1 Product Family Level Evaluation

Total Production Requirements Over The Planning Horizon (When Evaluated At Product Family Level)					
(1) Beginning Inventory Position	(2) Annual Selling Rate	(3) End of Planning Horizon Inventory Target	(4) Total Production Required Over Next 12 Months[a]	(5) Useable Inventory Over Planning Horizon[b]	(6) Excess Inventory Over Planning Horizon[c]
9,200	16,000	3,200	10,000	9,200	0

a) Calculation for column (4) = maximum of: [column (2) + column (3) – column (1), or 0]
b) Calculation for column (5) = minimum of: [column (1), or column (2) + column (3)]
c) Calculation for column (6) = maximum of: [column (1) - column (5), or 0]
*The planning horizon in this example is the next twelve months

Table 6.2 offers an analysis of the production requirements for this same product family calculated with one important difference. This figure displays the calculations that result when the beginning inventory of the product family is evaluated at the end item level, and then is aggregated to generate a beginning of period product family inventory number. Note that in this example, the product family contains five individual end items, and the respective total beginning inventory, forecast annual sales and end of planning horizon inventory target of these five items sum to the identical numbers shown for this product family in Table 6.1. A quick review of the *total* row for columns (5) through (7) in Table 6.2 reveals though that a different perspective of the inventory position of this product family emerges at the end item level. Specifically, analyzing the beginning inventory at this level shows that the vast majority of the total current inventory of this product family resides in end item 5. Further, the calculations show that 1,000 units of this end item's inventory are "excess over the planning horizon"; and therefore, only 8,200 units (column 6) of the inventory of the product family are "useable over the planning horizon". Assuming that there is no substitutability among end items within the product family, this means that a firm in this example must actually produce 11,000 units of the product family during the planning horizon, rather than the 10,000 units calculated by the analysis conducted strictly at the product family level.

Table 6.2 End Item Level Evaluation

Total Production Requirements Over The Planning Horizon (When Evaluated At End-Item Level)						
(1) End Item	(2) Beginning Inventory Position	(3) Annual Selling Rate	(4) End Of Planning Horizon Inventory Target	(5) Total Production Required Over Next 12 Months[a]	(6) Useable Inventory Over Planning Horizon[b]	(7) Excess Inventory Over Planning Horizon[c]
1	400	2,000	400	2,000	400	0
2	200	3,000	600	3,400	200	0
3	100	1,000	200	1,100	100	0
4	300	4,000	800	4,500	300	0
5	8,200	6,000	1,200	0	7,200	1,000
Total	**9,200**	**16,000**	**3,200**	**11,000**	**8,200**	**1,000**

a) Calculation for column (5) = maximum of: [column (3) + column (4) – column (2), or 0]
b) Calculation for column (6) = maximum of: [column (2), or column (3) + column (4)]
c) Calculation for column (7) = maximum of: [column (2) - column (6), or 0]
*The planning horizon in this example is the next twelve months

Figure 6.6 summarizes the difference that will result in this example from a production requirements forecast based upon a beginning inventory number generated at the product family level and one generated by first evaluating inventory at the end item level. To plan production requirements accurately, the firm must recognize that it has 1,000 units of inventory in item 5 (and therefore in item 5's product family) which it cannot utilize over the planning horizon. For production planning purposes, including these 1,000 units in the product family's beginning of period inventory figure will overstate the true level of inventory "useable over the planning horizon" for this product family. Thus, the quantity displayed in column (6) of Table 6.2 (8,200), rather than the quantity shown in column (5) of Table 6.1 (9,200), represents the proper amount to use as this family's "beginning of period inventory" in an annual production/distribution planning model. Obviously for other reasons (e.g., financial, special marketing programs, etc.), one must account for this product family's excess inventory over the planning horizon in other planning areas.

Inventory Evaluation At:	Production Requirement	Useable Inventory	Excess Inventory
Family Level	10,000	9,200	0
Item Level	11,000	8,200✓	1,000

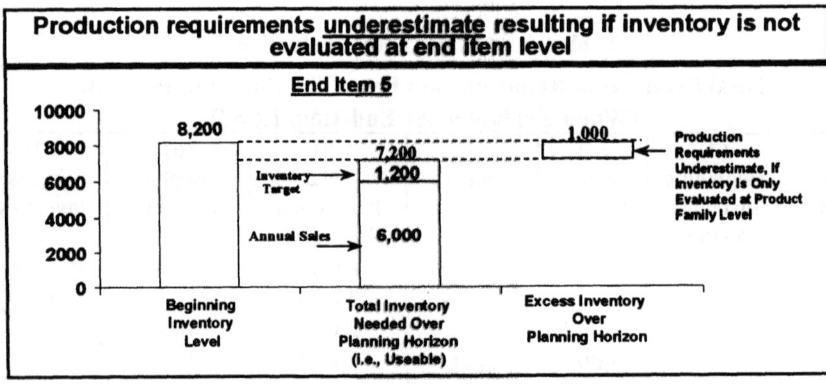

Figure 6.6 Product Family vs. End Item Level Evaluation

The preceding example illustrates that evaluating product family inventory positions (and production requirements) at the product family level makes it impossible to recognize if there is any excess or unuseable inventory over the planning horizon in any of the end items within the product family. By using the hierarchical approach presented here (*i.e.*, evaluating product family inventories at the more disaggregated item level, and then summing end item results to obtain beginning of period product family inventory data), one can accurately evaluate a product family's current level of "useable inventory over the planning horizon." The use of this algorithm or a similar approach assures that a firm will not, as a result of a mis-evaluation of the utility of its existing inventory, underestimate the true level of capacity that it will require over a planning horizon.

A Simple Algorithm To Determine Useable And Excess Inventory Over A Planning Horizon

The example shown in Figures 6.5 and 6.6, and Tables 6.1 and 6.2 demonstrates the calculation of "useable" and "excess" inventory over the planning horizon for one product family in one location. In the real world of multi-location manufacturing and distribution networks, the determination of proper useable and excess inventory data for product families becomes more complex. We close this section with an appendix that presents an algorithm which extends the same basic logic used for the one product family, one location example to a multi-location, multi-echelon network.

Appendix 6.3 A Multi-Echelon End Item Inventory Evaluation Algorithm

The following algorithm classifies the inventory of each end item at each location (or groups of locations) into two categories: (1) useable over the planning horizon and (2) excess over the planning horizon. Although we present this algorithm in the context of a simple two echelon illustration, one can easily extend the algorithm's basic logic to address networks with more than two echelons, and with locations within echelons aggregated or disaggregated as desired.

For illustrative purposes, we formulate this algorithm assuming that it will be applied to a distribution network consisting of two plants and N stores (or small warehouses). Each of these (2 + N) locations is assumed to stock finished goods inventory and have inventory targets (*i.e.*, objectives) for the end items which it stocks (sku's). For purposes of simplicity, we assume that inventory objectives remain constant for sku's over the planning horizon, and we will use the term "end-of-period inventory objective" to refer to these targets. (Note that with minor modifications, this algorithm can be altered to accommodate inventory objectives changing during a planning horizon.) It is also assumed for simplicity that there exists no substitutability among end items and that forecasts are accurate. The N stores are assumed to place replenishment orders upon the two plants, and the two plants maintain inventories (sku's) to supply the N stores upon demand. The N stores similarly maintain inventories to supply customer demands immediately from inventory (*i.e.*, the distribution network operates in a make-to-stock environment.) Given this assumed network, we now present the following end item inventory evaluation algorithm.

Table 6.3 Notation For End Item Inventory Evaluation Algorithm

Let A and B represent the two plants.
N is the number of stores (or field warehouses) on the firm's distribution network.
TAI^i is the actual total company-wide beginning-of-period inventory level of the ith end item. (Beginning of period refers to beginning of the planning horizon.)
MIR^i is the maximum or total quantity of beginning-of-period inventory of the ith end item which the firm could theoretically utilize over the entire planning horizon. This is defined as the sum of the total company-wide inventory objective for the ith end item and the total sales forecast for the ith end item over the planning horizon.
$AI^i(A)$, $AI^i(B)$, $AI^i(N)$ is the actual beginning-of-period inventory of the *ith* end item at plant A, plant B, and the N sales stores, respectively. N can also be thought of as a fictitious location representing the aggregation or sum of all sales stores.
$IO^i(A)$, $IO^i(B)$, $IO^i(N)$ represent the inventory objective for the ith end item at plant A, plant B, and the N stores respectively.
$EIH^i(A)$, $EIH^i(B)$, $EIH^i(N)$ represent the quantities or portions of the inventory of the ith end item at plant A, plant B, and the N stores respectively which the firm theoretically will not be able to utilize over the planning horizon. These inventory quantities are defined as "excess over the planning horizon."
$UIH^i(A)$, $UIH^i(B)$, $UIH^i(N)$ represent the quantities or portions of the inventory of the ith end item at plant A, plant B, and N stores respectively which the firm theoretically will be able to utilize (to satisfy a demand or inventory need at some location) over the planning horizon. These inventory quantities are defined as "useable over the planning horizon."
$A\%D^i$ is the average percentage of the total demand for the ith item which plant A fills (of the combined total demand for the ith item placed upon plant A and Plant B by the N stores). This percentage is determined based upon recent demand history for a determined representative period (e.g. the previous six months).
$B\%D^i$ is the average percentage of the total demand for the ith item which plant B fills (of the combined total demand for the ith item placed upon plant A and plant B by the N stores).
$AMAX^i = (A\%D^i) * (MIR^i)$ is a computed value (to be explained later).
$BMAX^i = (B\%D^i) * (MIR^i)$.

Figure 6.7 details the logic used to classify the inventory of each end item of plants *A*, *B*, and the *N* field warehouses or stores into "useable" and "excess" over the planning horizon. Each of the steps shown in Figure 6.7 will now be described.

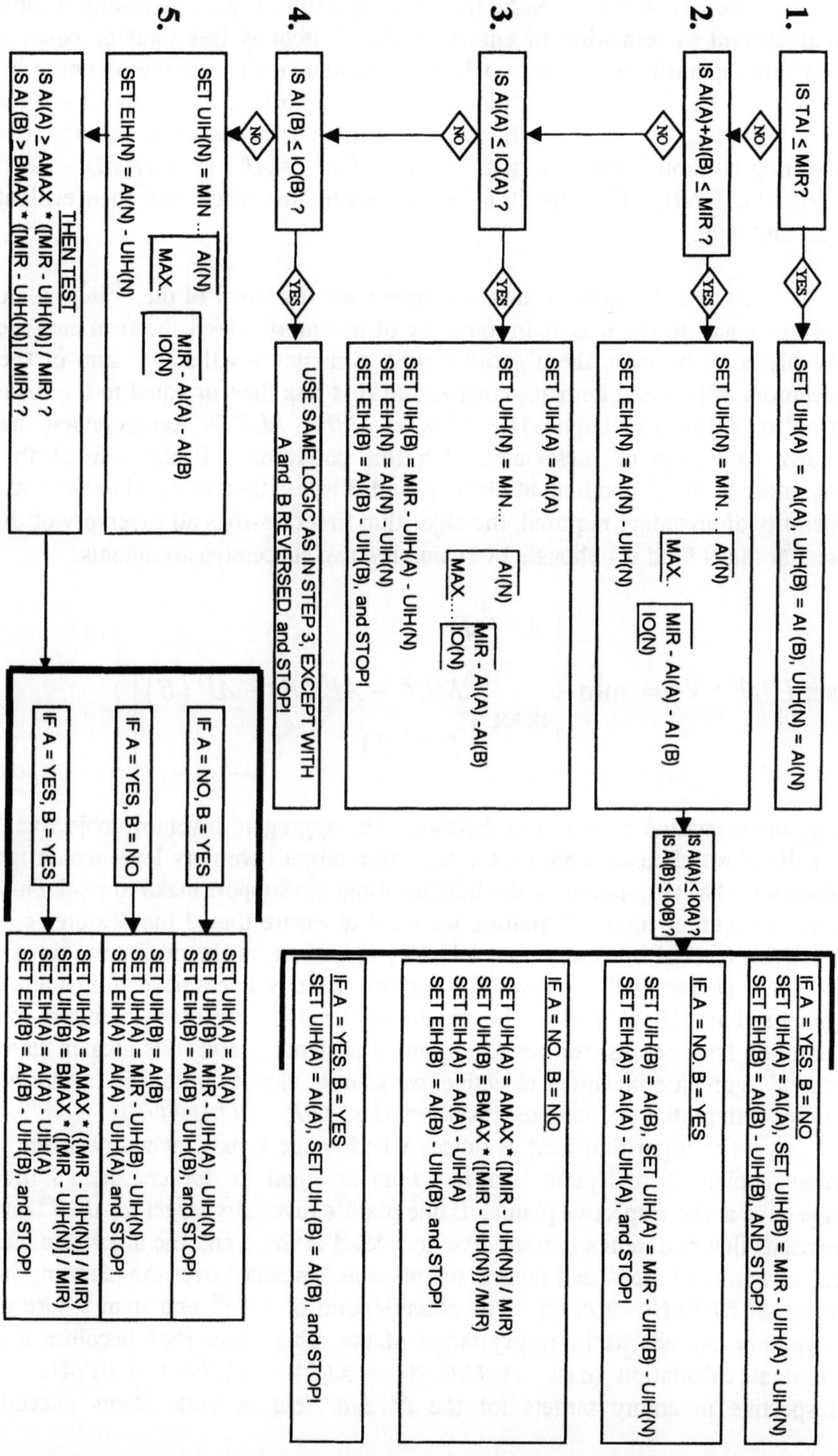

Superscript *i* is omitted in this figure.

Figure 6.7 End Item Inventory Evaluation Algorithm

Step 1. For each end-item i, the algorithm begins by testing whether the total current system-wide inventory of the ith item is less than or equal to the maximum quantity of inventory which the firm can use over the planning horizon (is $TAI^i \leq MIR^i$). If this condition is true, the algorithm classifies all inventory of the ith end item at plants A, B, and the N field warehouses as "useable over the planning horizon," and then stops. (Set $UIH^i(A) = AI^i(A)$; $UIH^i(B) = AI^i(B)$; set $UIH^i(N) = AI^i(N)$). The algorithm would record this result and then evaluate the next end item.)

Step 2. If the total current system-wide inventory of the ith item is not less than or equal to the maximum quantity of inventory which the firm can use over the planning horizon, the algorithm next evaluates whether the sum of the total inventory of this end item at plants A and B is less than or equal to the maximum quantity of inventory required (is $AI^i(A) + AI^i(B) \leq MIR^i$?). Let us review the logic shown in Figure 6.7 associated with this condition. If the sum of the total inventory of the ith end item at plants A and B is less than or equal to the maximum quantity of inventory required, the algorithm first classifies all inventory of this end item at the N field warehouses by evaluating two successive arguments:

$$\text{set } UIH^i(N) = \min \begin{Bmatrix} AI^i(N) \\ \max \begin{Bmatrix} MIR^i - AI^i(A) - AI^i(B) \\ IO^i(N) \end{Bmatrix} \end{Bmatrix}$$

The interpretation of this is as follows. The aggregate inventory objective $IO^i(N)$ for the N warehouses represents a minimum target inventory level which the firm always wishes to maintain at the field locations (to support make to stock customer-service requirements). Therefore, we want to assure that if the N stores currently have this level of inventory [*i.e.*, $IO^i(N)$] or greater, at the minimum, the quantity of $IO^i(N)$ is classified as "useable." Further, if the N stores have an inventory level greater than $IO^i(N)$, and if the maximum level of inventory required over the planning horizon, less the actual inventory at plants A and B, is greater than the N stores' aggregate inventory objective, we want to classify all of the inventory equal to this differential as "useable" inventory (*i.e.*, $MIR^i - AI^i(A) - AI^i(B)$).

The algorithm next evaluates whether the actual inventory of the ith end item at plant A and plant B is less than or equal to this end item's inventory objective at the respective plants. If one plant's inventory target for the ith end item exceeds that end item's current inventory level at the plant, the algorithm classifies all inventory of the ith end item at the plant as "useable" over the planning horizon (e.g. set $UIH^i(A) = AI^i(A)$). The classification of the ith end item whose current inventory exceeds its inventory target at the other plant then becomes a simple residual calculation (e.g., set $UIH^i(B) = MIR^i - UIH^i(N) - UIH^i(A)$). If the respective inventory targets for the ith end item at both plants exceed their respective current inventory levels of this item, the algorithm denotes all inventory of this item at both plants as "useable."

In the case where the inventory of the ith end item at each plant exceeds its respective inventory objective, the algorithm employs two calculated parameter values, $AMAX^i$ and $BMAX^i$ to determine the useable inventory of this end item. Briefly, $AMAX^i$ and $BMAX^i$ represent fictitious allocations between plant A and plant B of 100% of the maximum inventory required over the planning horizon. We base these allocations upon the historical average percentage of demand placed upon both plants for the ith end item by the N stores. Multiplying $AMAX^i$ and $BMAX^I$ by the ratio of (maximum inventory required less the calculated useable inventory at the N stores) to (maximum inventory required) yields the useable inventory of the ith end item at plants A and B, respectively.

Step 3. If the total inventory of the ith item at plants A and B is not less than or equal to the maximum quantity of inventory required, the algorithm next evaluates whether the total inventory of this end item at plant A is less than or equal to the inventory objective of this end item at plant A (is $AI^i(A) \leq IO^i(A)$?). If this condition is true, all inventory at plant A is classified as "useable over the planning horizon" [set $UIH^i(A) = AI^i(A)$]. The algorithm next categorizes the inventory at the N field warehouses by the identical process previously outlined in step 2. Finally, we obtain the "useable" inventory at plant B as a residual value (set $UIH^i(B)=MIR^i-UIH^i(N)-UIH^i(A)$).

Step 4. If the total inventory of the ith end item at plant A is not less than or equal to the inventory objective of this end item at plant A, the algorithm then tests whether plant B's inventory of this end item is less than or equal to its inventory objective. If this condition is true, the algorithm uses the same logic reviewed in step 3 (except that it is structured based upon plant B's inventory).

Step 5. If the algorithm reaches this step, it has already determined that the combined inventory of plants A and B exceeds the maximum required over the planning horizon for the ith end item, and that each plant's inventory of the ith end item exceeds its respective plant inventory objective. In this case, we first determine the "useable" inventory at the N field warehouses as previously described. The algorithm then tests whether the respective actual inventories at plant A and B are greater than or equal to the previously described values of $(AMAX^i * ((MIR^i - UIH^i(N))/MIR^i))$ and $(BMAX^i * ((MIR^i - UIH^i(N))/MIR^i))$. As Figure 6.9 indicates, the classification of inventories at A and B are obtained based upon the results of these two tests.

From this review, it should be clear that the inventory analysis/classification scheme of this algorithm has a relative bias towards categorizing inventories at the central stocking location(s) of the distribution network (in this example, the plants) as "useable over the planning horizon." This relative bias results from the fact that the calculated value of $UIH^i(N)$ can only exceed the inventory objective of the ith end item $(IO^i(N))$ if the value $[MIR^i - AI^i(A) - AI^i(B)]$ exceeds $IO^i(N)$. We present this approach because practical experience suggests that planning to utilize "useable" inventory dispersed among many locations on a distribution network has a higher risk than planning to utilize "useable" inventory already consolidated in a central location. Thus, for any end items which have a system-wide inventory level greater than the maximum required over the planning horizon (MIR^i), we have constructed the algorithm such that it would designate the central location's inventory (greater than target levels) as "useable" before it would designate field inventory (greater than target levels) as "useable."

However, one can easily modify this algorithm to designate field inventory (greater than target levels) as "useable" before it would designate central location inventory (greater than target levels) as "useable". Further, using the general approach embodied in this algorithm, one can create other similar algorithms to: (1) address any network configuration, regardless of complexity, and, (2) to address a firm's general philosophy towards the utility of inventory over a planning horizon.

Finalizing Beginning-Of-Period Product-Family Inventory Data

Once we have run the inventory-evaluation algorithm for all end items, we can then calculate the proper inventory levels to use as the beginning inventory data for each product family at each location defined in our long-run production/distribution planning models. For each product family at each location, we simply sum the inventory defined as "useable over the planning horizon" (UIH^i) for each end item within that product family. These sums provide the proper beginning-of-period inventory quantity for each product family at each location.

6.4 Hierarchical Approaches For Developing Tactical And Operational Planning Forecasts

This section turns to the topic of forecasting and in particular considers several illustrative approaches that use hierarchical planning principles to generate tactical and operational planning forecasts. We begin by posing a simple description of what hierarchical forecasting entails. Hierarchical forecasting involves employing a top-down and a bottom-up methodology to forecast individual sku's, end items, and/or groups of end items (*i.e.*, product families and higher level groupings of items and families). Figure 6.8 offers an illustration of top-down and bottom-up forecasting. In top-down forecasting, some methodology must be employed to disaggregate higher level forecasts into lower level forecasts. Methodologies employed to accomplish this include: (1) using the historical ratio of individual items' sales or demand to the sales of the group of items of which each item is a member, and (2), when appropriate, apportioning a product group forecast equally amongst all the items which compose the group. Bottom-up forecasting consists simply of summing up individual forecasts to obtain a higher level aggregate forecast, as Figure 6.8 illustrates. In the forecasting literature, the terms "top-down" forecasting and "bottom-up" forecasting are often used interchangeably with the term "hierarchical" forecasting, and we will not attempt to dispute this terminology. However, later in this section, we will present a brief example of a forecasting application that utilizes both top-down and bottom-up forecasting methodologies together with a feedback loop in order to develop a final forecast. As a minor aside comment, one can posit the argument that a "true" hierarchical forecasting methodology must include some sort of feedback mechanism or loop, just as other hierarchical planning approaches such as hierarchical production planning do. However, the important issue in forecasting is simply what approach and algorithm will produce the most accurate forecasts.

Figure 6.4 illustrated a typical "product" hierarchy that a firm might utilize to define and aggregate its product line. (Recall that the four levels from bottom to top consisted of item, product family, product line and product division.) Hierarchical forecasting methods can work well to provide forecasts at each level of such a product hierarchy.

It is important to note that there exist numerous dimensions in addition to "product" that firms frequently apply hierarchical methods to forecast. These dimensions include:

- geographic (e.g., countries, states,...)
- sales organization (e.g., regions, districts, territories,....)
- sales channels (e.g., mass merchandisers, wholesalers,....)
- company organization (e.g., company, business unit, sub-unit,...)
- planning horizon/time period (e.g., years, quarters, months, ...)
- network locations (e.g., plants, DC's,...)
- customers (e.g., customer, customer location,...)

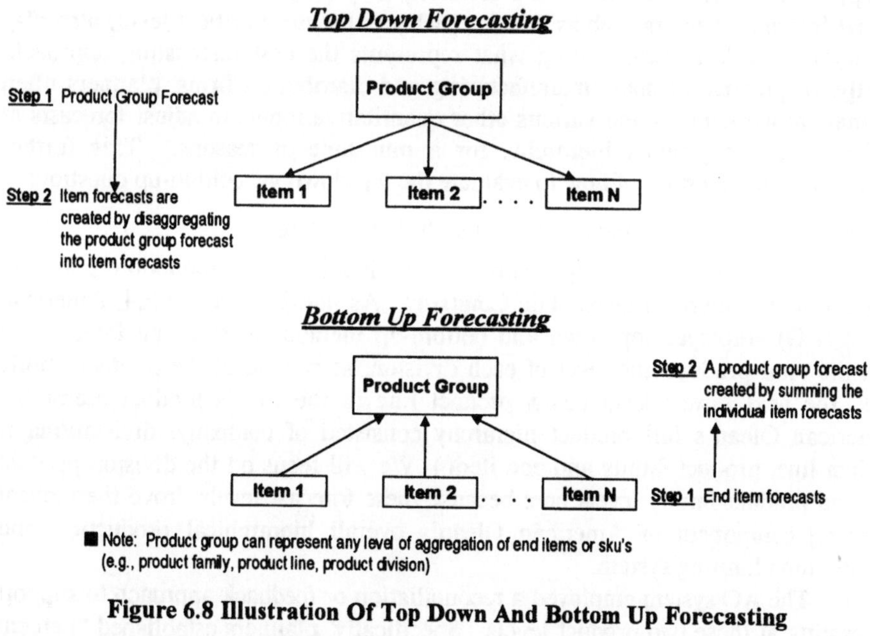

Figure 6.8 Illustration Of Top Down And Bottom Up Forecasting

Regardless of the "dimensional" approach, ultimately a firm wishes to produce the most accurate forecast possible in terms of sales dollars and unit sales. Numerous factors can dictate the dimension or dimensions a firm will employ for their forecasting purposes; however, which dimension yields the best forecasts, and which dimension best aligns with the firm's organizational structure represent two key considerations.

The forecasting literature in such journals as the International Journal of Forecasting and The Journal of Business Forecasting is replete with research articles comparing the merits of bottom-up and top-down forecasting methods. The interested reader is referred to these two journals and in particular to the following articles which provide a sampling of this research (Dangerfield and Morris, 1992; Gordon, Morris and Dangerfield, 1997; Kahn, 1998; and Lapide, 1998). For our purposes in this subsection, we simply note that:

1. The particular application will determine whether bottom-up or top-down represents the most effective approach (if one must choose between one method or the other), and
2. a number of researchers have suggested that a mix of both methods may often represent the best approach (see e.g., Kahn, 1998).

It is also important to note that much of the debate on the merits of top-down vs. bottom-up forecasting focuses on what method produces the most accurate forecast at the lowest level of forecasting (e.g., at the item or sku level). Depending upon the application, the lowest level of a hierarchy may or may not represent the most critical forecast. Factors such as this (*i.e.*, what is the most critical level) also play an important role in determining what represents the best forecasting approach. Finally, in practice at many manufacturing and distribution firms, planners often use manual judgements and various other quantitative inputs to adjust forecasts at all levels of the product hierarchy, for a multitude of reasons. This further complicates the question of how to evaluate the top-down vs. bottom-up question.

Utilizing Hierarchical Planning Principles In Forecasting

In this subsection, we briefly return to the hierarchical production planning application of American Olean Tile Company. As noted in section 2.4, American Olean (AO) employed top-down and bottom-up methods to generate forecasts at both the total product line level of each division, as well as at the product family level, the next lowest level below product line in the firm's product hierarchy. (American Olean's full product hierarchy consisted of company, division/major product line, product family and end item.) We will focus on the division product line and product family levels here because these forecast levels drove the tactical planning component of American Olean's overall hierarchical production and distribution planning system.

The AO system employed a reconciliation or feedback approach to support forecasting at these two product levels. Specifically, planners established "percent and/or absolute difference" tolerances for purposes of monitoring the top-down and bottom-up generated forecasts at both (1) the division level and (2) the individual product family level. If AO's hierarchical forecasting process resulted in a forecast that exceeded the set tolerances for any of the following cases, this triggered a review to investigate and reconcile the difference in forecasts. Planners set tolerances to monitor:

1. forecasts for the *total product line* of a division yielded by:
   a) the top-down division level forecast, and
   b) the sum of the bottom-up product family level forecasts generated for each product family in a product line, and by,

2. forecasts at the *product family level* of a division yielded by:
   a) the top-down division level forecast dissaggregated into individual product family forecasts on the basis of the historical ratios of the individual families sales to the total division's product line sales, and
   b) the individual product family level forecasts generated for each product family in a product line as part of the bottom-up approach.

This hierarchical forecasting process improved the overall forecasting accuracy of the firm, raised the level of corporate attention devoted to the tactical forecasting process, and contributed to the overall benefits of the firm's hierarchical production planning implementation, as previously outlined in section 2.4. An additional benefit of this approach is that it assured that forecasts created by different planners at different levels of aggregation were aligned.[1] The actual resolution process consisted of planners responsible for the different levels of forecasts working together to first understand, and then resolve significant differences in the forecasts generated by the top-down and bottom-up approach. Typical steps in the resolution process included obtaining appropriate field and headquarters input (e.g., from the sales organization), as well as reviewing the forecast algorithms and time series data.

A hierarchical forecasting approach that utilizes both top-down and bottom-up forecasts, and which further attempts to assure alignment between forecasts has both potential advantages, disadvantages and requirements. Such an approach does facilitate the use and alignment of multiple forecasting methodologies, each of which may represent the best algorithm to employ at a particular level in the product hierarchy. For example, causal forecast methods such as regression and econometrics may in some cases yield the best results at an overall firm or major division level. On the other hand, more traditional time-series forecasting methods such as exponential smoothing may represent the appropriate forecasting algorithm to employ at the product family level. The hierarchical approach just described allows a firm to effectively use different methods in a complementary fashion.

[1] See Liberatore and Miller, 1985; and De Matta and Miller, 1993 for additional discussion of these benefits.

This type of multi-level top-down and bottom-up forecasting system often requires more work and support, however, than does a simpler forecasting approach (e.g., a simple top-down approach). In addition to requiring the development of as many as double the number of forecasts, the forecast reconciliation process generally necessitates the participation of a number of planners and other members of a firm. Automation of much of the forecasting and monitoring process can reduce the level of ongoing work required (e.g., the effort to produce quarterly or monthly updates of forecasts). Nevertheless, a firm should evaluate the cost of the extra resources required to support a multi-level top-down and bottom-up forecasting system versus the potential or actual improvement in accuracy that such a system may yield. This trade-off will vary depending upon the particular firm and forecasting application.

Finally, consideration of using a dual top-down and bottom-up forecasting methodology to generate two different forecasts for the same level (e.g., the product family level) only has merit in forecasting environments that allow the possibility of developing acceptably accurate forecasts by both methodologies. For example, as a general rule, one should only use a top-down approach to generate forecasts at a "lower" level in the product hierarchy, if the sub-groupings (or items) at the lower level follow the same basic demand pattern as the product grouping at the "higher" level. Figure 6.9 illustrates two simple examples where a top-down forecasting method could possibly work (case A), and where such an approach would not yield good lower level forecasts (case B).

In case A, each of the items in the product family exhibit a similar seasonal demand pattern. The individual items may have significantly different absolute sales levels; however, the similarity of their relative demand patterns makes it plausible to consider a top-down forecasting approach. In contrast, in case B we observe that the individual items display dramatically different demand patterns. A top-down forecasting approach to these items would clearly yield poor forecasts at the item level. Thus, in case B, a multi-level top-down and bottom-up methodology where planners compared item forecasts generated by both a top-down forecast and a bottom-up forecast would not represent a good approach. It still remains critical in such cases, however, to assure alignment between forecasts at different levels of the product hierarchy (*i.e.*, that the final end item forecasts sum up to the final product family forecasts and so on).

We close this section on hierarchical forecasting by briefly noting that there exist numerous other forms of forecasting that have different hierarchical dimensions to their approach. For example, a hierarchical forecasting process becoming increasingly popular in practice consists of integrating customer forecasts (for a manufacturer's products) directly into a manufacturer's forecasting process. Briefly, the hierarchical dimensions in this case include developing forecasts by integrating information collaboratively across multiple echelons of the supply chain, comparing customer forecasts against a manufacturer's own forecasts and so on. This subject represents a major forecast topic in itself, and for purposes of this discussion, we simply note that it represents an excellent example of a different form of hierarchical forecast planning. In the next section, we will review how hierarchical forecasting and inventory management techniques work together in

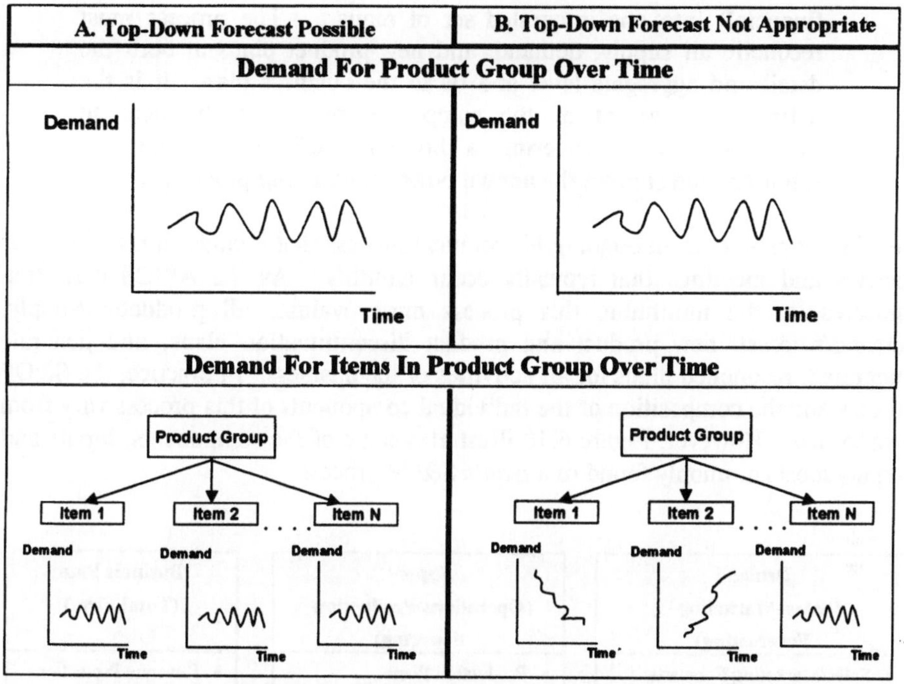

Figure 6.9 Illustration Of When Top Down Forecasting Is Applicable

support of another popular planning methodology, the Sales And Operations Planning process.

6.5 Sales And Operations Planning (S&OP)

In recent years, a planning approach commonly known as the "Sales And Operations Planning" process (see e.g., Ling and Goddard, 1988) has enjoyed increased popularity. Many firms have implemented the S&OP process and consider it an integral component of effective supply chain planning and operations. S&OP methodologies and implementations often vary significantly by firm. However, the various permutations of the S&OP methodology found in practice share the common characteristic that they rely heavily on a hierarchy of forecast, inventory and production planning activities and data. In this section, we will explore some of the hierarchical dimensions of the S&OP process.

The American Production And Inventory Control Society (APICS) defines sales and operations planning as follows:

> A process that provides management the ability to strategically direct its businesses to achieve competitive advantage on a continuous basis by integrating customer-focused marketing plans for new and existing products with the management of the supply chain. The process brings together all the plans for the business (sales, marketing, development, manufacturing, sourcing, and

> financial) into one integrated set of plans… The process must reconcile all supply, demand, and new-product plans at both the detail and aggregate level and tie to the business plan. It is the definitive statement of the company's plans for the near to intermediate term covering a horizon sufficient to plan for resources and support the annual business planning process.[2]

The S&OP process is an ongoing hierarchical process that includes a set of regular reviews and meetings that typically occur monthly. As the APICS definition indicates, at the minimum, this process must evaluate all production/supply, demand/forecast, new product and product discontinuation plans; and generate integrated, reconciled plans for all activities of the business. In practice, the S&OP process and the composition of the individual components of this process vary from firm to firm. However, Figure 6.10 illustrates some of the components, inputs and outputs most commonly found in a firm's S&OP process.

Demand (Sales–Marketing–Forecasting)	Supply (Operations-Production-Sourcing)	Business Plan (Total Firm)
• Sales/Marketing Forecasts • Inventory Analysis Of Customers • New Product Introductions • Product Discontinuations • Advertising And Promotion Plans • Product Return Projections • Performance Measure Reviews • Sales, Marketing And Distribution Cost Projections • Sensitivity Analysis	• Production Plans • Inventory Plans • Customer Service Fill Rate Objectives • Capacity Utilization Projections • Capacity Contingency Plans • Performance Measure Reviews • Manufacturing and Purchasing Cost Projections • Sensitivity Analysis	• Revenue Projections • Total Cost Projections • Profit Projections • Sensitivity Analysis

Figure 6.10 Illustrative Components Of S&OP Process

It is important to note that particularly in the area of strategic planning, other components not depicted in Figure 6.10 such as long range competitive analyses, industry forecasts, long-range industry capacity level projections, firm manufacturing capacity expansion and/or contraction plans, and so on would also play a prominent role in the process. Additionally, note that "sensitivity analysis" appears under the demand side, the supply side and the overall business plan representations of the S&OP process. Performing sensitivity analysis on key plans facilitates the understanding and management of risk in all aspects of the firm's planning process. Thus, one cannot overemphasize the importance of sensitivity

---

[2] See APICS Dictionary, 1995

analysis in a firm's S&OP. Before considering the increased complexities that a "global" planning process adds to the basic S&OP process, we close this discussion on the S&OP with a generic illustration of how the components depicted in Figure 6.10 fit together in an integrated business process.

A firm's actual S&OP schedule will require much more detail than shown in Figure 6.11. It will include definitions and monthly schedules for each activity performed. For example, the schedule should define the business day of the month when the firm's updated forecast is reviewed and finalized, the day of the month on which the updated production plan is reviewed and finalized, etc. However, this figure offers a general description of how individual S&OP components must fit together to generate an integrated business plan.

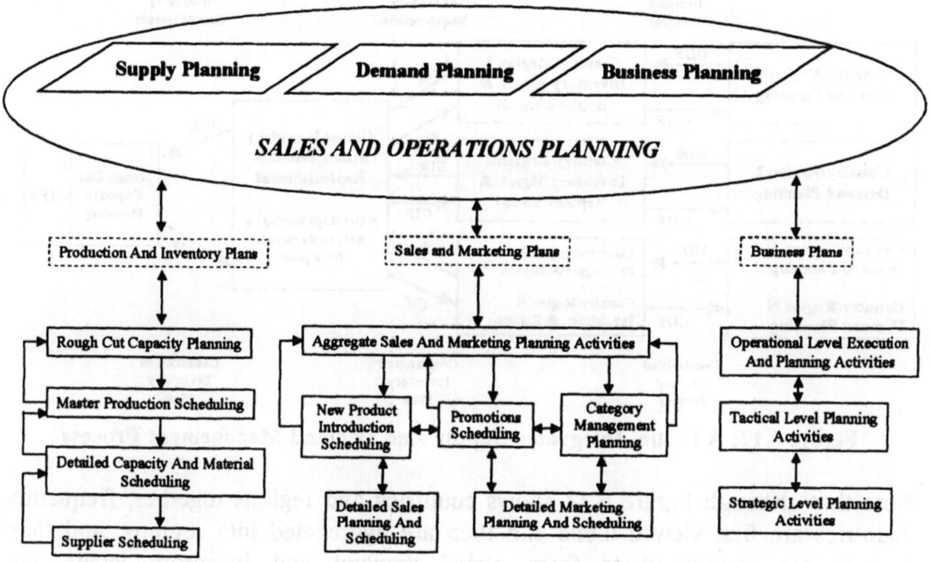

[Note: The planning hierarchy is shown in reverse (i.e. inverted) order here to match the S&OP's focus on the operational level]

Figure 6.11: Integrating S&OP Into An Overall Business Planning Process

The Hierarchical Characteristics Of A Global S&OP Process

Conducting an S&OP process at a global or even "multiple country" level raises additional hierarchical planning issues to address. At the global level, the complexity of the planning process escalates rapidly; the ability to maintain a coordinated plan becomes increasingly challenging as the number of, and distance between geographic regions rises; and such issues as the timing and the flow of information between disparate information systems located around the globe present major hurdles to overcome in operating an effective global S&OP. It is beyond the scope of this book to tackle these issues in detail, and the interested reader is referred to Ling and Goddard (1988) for detailed presentations on this subject. However, a quick review of one approach to the global supply and demand planning subcomponent of the S&OP process will serve to illustrate the hierarchical nature and complexity of the global S&OP process.

Figure 6.12 depicts a methodology that facilitates integrated supply and demand planning across countries and regions for large international firms. The process shown in Figure 6.12 can support a network that includes country and regional plants (*i.e.*, a plant that serves only one country or one region of the world), as well as global plants. We define a "true" global plant in this example as a plant that supplies the total demand placed on the firm, in all countries served by the firm, for one or more products produced by the particular plant. That is, the plant represents the sole or primary supply source to all countries served by a firm, for one or more products.

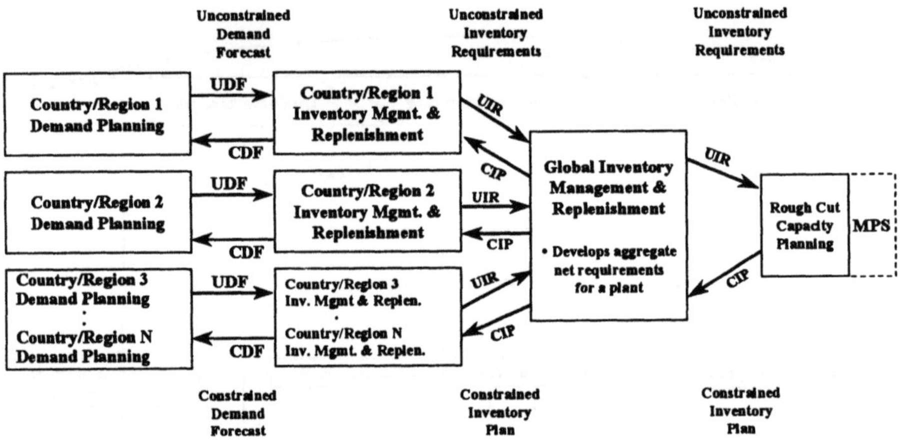

Figure 6.12: A Global Integrated Supply And Demand Management Process

Note that although Figure 6.12 shows countries and regions together, frequently countries are first viewed alone and then are aggregated into regions, and then regions are aggregated to form global demand and inventory levels and requirements. In a similar fashion, the "disaggregation process" often flows from the global level to the regional level to the country level. For simplicity, we have not shown this additional level of detail.

The monthly global S&OP process portrayed in Figure 6.12 begins with a forecast by product family at the country level, or perhaps at the regional level (in the cases of small "sales" volume countries located in close geographic proximity to each other). Note that the initial forecast in this illustration is an "unconstrained" demand forecast, meaning that production capacity and inventory constraints are ignored at this stage of the process (*i.e.*, the focus is on developing the best forecast of demand, regardless of expected product availability). In the next stage of this process, planners use tools such as distribution resource planning systems to create unconstrained inventory requirements from the unconstrained demand forecasts, again at the product family level. This activity entails appropriately adding end-of-period inventory targets at the country/region level to the unconstrained forecasts and subtracting the beginning of period inventory levels. To understand the potential complexity of this process, envision this activity occurring on a monthly basis across a global network with thousands of sku's and numerous stocking locations. Further, imagine the systems and planning capabilities required to implement a true hierarchical planning process that utilized an algorithm similar to that described in section 6.3. (Recall the method presented to evaluate beginning of period end item and family inventory levels.) This can represent a very challenging system to implement in practice.

The next stage of this global planning process consists of translating country/regional unconstrained inventory requirements into global unconstrained inventory requirements. This requires that planners aggregate all regional/country unconstrained inventory requirements, and if appropriate, factor in any "global beginning and ending inventory period targets" not already accounted for at previous planning levels. For example, planners might factor existing inventories and ending inventory targets for finished goods at the global plant at this stage of planning. Once forecasting, inventory management and demand management planners[3] have determined the unconstrained inventory requirements at the global level, a manufacturing response to these requirements must be developed. As Figure 6.12 illustrates, planners must utilize production planning techniques such as rough cut capacity planning and master production scheduling to formulate this response.[4] This process evaluates whether sufficient manufacturing capacity exists to meet the global unconstrained inventory requirements. Next, the updated global manufacturing plan returns to the regional/country level indicating the "constrained" inventory plan. This plan describes what and when manufacturing proposes to deliver finished goods inventory to regions/countries. If sufficient manufacturing capacity exists to satisfy the original unconstrained inventory requirements, the constrained inventory plan will reflect the same products and quantities originally requested. In this case, as shown in Figure 6.12, an updated "constrained demand forecast" then flows from the region/country inventory management function to the region/country demand planning function. This step basically confirms that the original unconstrained demand forecast at the region/country level can be met (*i.e.*, adequate inventory will be available to meet the forecast), and these demand forecasts are then considered final for the current month's global S&OP. This allows the firm to finalize the business plan elements of the S&OP such as the rolling revenue and profit projections, shown in Figure 6.10. At this point, the firm will have completed its closed loop monthly global S&OP process.

However, if at the global manufacturing planning stage, planners determined that the global unconstrained inventory requirements exceeded the available production capacity, a variation on this process takes place. Specifically,

[3] There are a number of effective alternative organizational designs that firms employ to carry out the demand and supply management process. In practice, functional responsibility for activities such as forecasting, inventory and demand management, manufacturing planning and so on vary widely from firm to firm. Further, in some firms, single planning groups may have responsibility for a wide range of functions. Therefore, when discussing the flow of planning information from one activity to the next in this section (and this book), we are not attempting to imply that a separate planning entity must exist to handle each individual planning function.

[4] Refer to Chapters 2 and 3 for a review of the full-set of production planning techniques that can be used to generate the manufacturing response.

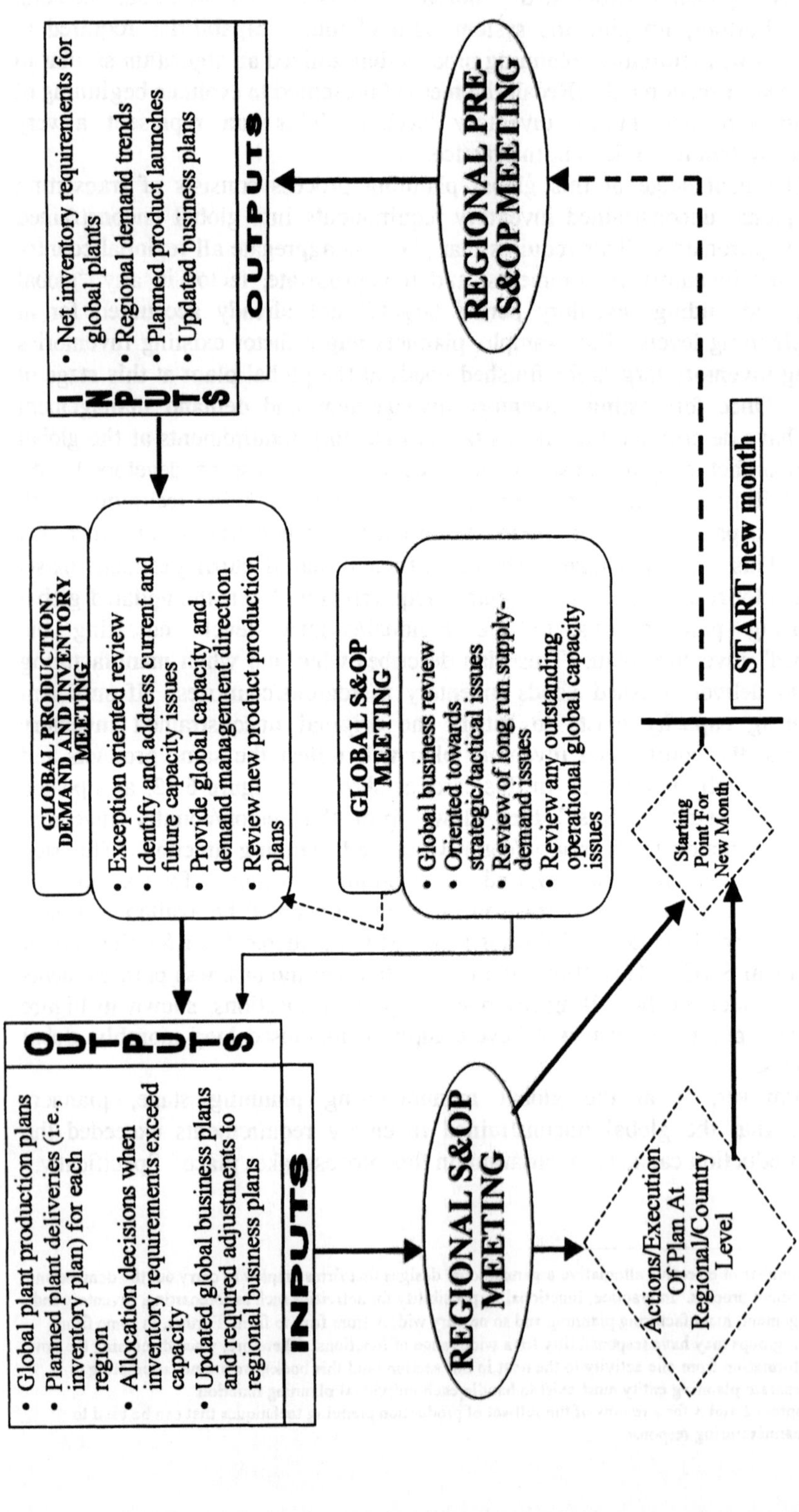

Figure 6.13 Illustrative Regional/Global S&OP Monthly Meetings

assuming that the capacity shortfall cannot be made up in the short term, the constrained inventory plan returned to the inventory management/demand management functions in this case will represent manufacturing's best attempt to satisfy as much of the original unconstrained global inventory requirements as possible, but not all of these requirements. This shortfall or gap will trigger an iterative process in which such functions such as marketing, sales and manufacturing determine the best course of action, given that the firm cannot fully satisfy forecast demand in the short run. Naturally, such factors as revenue and profit implications, longer term market share ramifications, etc. will play into the decision as to how to most effectively utilize available capacity. This iterative process ultimately will result in an agreed upon constrained inventory plan (different than the original unconstrained inventory requirements), and ultimately in a final "constrained demand (sales) forecast" at the region/country level.

The illustrative hierarchical supply and demand planning system that we have just described in Figure 6.12 entails numerous steps and can require several iterations in cases of capacity shortfalls, revenue and profit shortfalls, etc. When one considers that many practitioners suggest the global S&OP process should occur monthly to assure well-coordinated, integrated operations (see e.g., Ling and Goddard, 1988), the challenge of conducting such a process on a global scale within the time constraints available becomes clear. Further, while the system depicted in Figure 6.12 represents the core set of supply/demand planning activities of the monthly global S&OP, it also represents only part of the overall S&OP process. The activities detailed in Figure 6.12 primarily address the operational level monthly supply and demand planning process. In addition to these activities, operational financial and business plans must be integrated into this process, as well as longer-term tactical and strategic issues.

Figure 6.13 displays an illustrative global S&OP schedule of monthly meetings, each of which provides inputs and/or outputs to a different level in the planning process. Again for simplicity, we present regional/country S&OP's on the same level, although we note again that country level activities either precede or occur in parallel with regional activities, depending upon the relative size and importance of countries and regions in a particular firm. As exhibited in Figure 6.13, regional/country "pre-S&OP" meetings precede the monthly "global S&OP" meeting(s). At these regional pre-S&OP meetings, the key activities and plans of the S&OP depicted in Figures 6.10 and 6.11; namely, demand (sales), supply (operations) and the business plan are reviewed. The "global" unconstrained inventory requirements depicted in Figure 6.12 would represent just one area reviewed at this meeting.

The outputs of the regional pre-S&OP meetings become the inputs to the global S&OP review meeting(s). The global S&OP meeting(s) is a high level, exception-oriented review which addresses global operational supply and demand issues (using inputs from the process depicted in Figure 6.12), operational global business issues, as well as more tactical and strategic issues. Typical questions tackled at this global review include the following:

- What is the best long term network infrastructure strategy?
- How should long term capacity shortages be resolved?
- How do planned new product introductions impact global capacity requirements?
- How should trade-offs be made when multiple countries demands for capacity conflict?
- What is the best allocation of capacity from a global business point of view?
- How should global inventory imbalances or shortages be addressed?
- How should forecasts be adjusted based on current business conditions?
- What are appropriate global service and inventory targets?
- What are risks to the global business plan?
- What are the business conditions at the regional levels?
- What are the best hedge or gap alternative plans?
- How should global and country financial targets be adjusted based on current business conditions?

Decisions made at the global S&OP on the preceding questions become inputs to the monthly S&OP meetings that take place at each region/country in a firm's hierarchy. In particular, decisions made regarding capacity allocations and the final constrained global inventory plan allow regions and/or countries to then finalize their monthly S&OP plans. For example, if an inventory shortage exists in a particular global product family, the final constrained global inventory plan allows regional decision-makers to then determine the allocation of inventory of the product family to the countries that compose a region. At this point, planners can update (*i.e.*, finalize) the constrained demand (sales) forecast for this product family at the country/region level. This represents one small example of the type of closed-loop planning decision that must originate at the country/region level, flow up to the global decision-making level, and then flow back down to the regional/country level. Clearly, given the importance of speed of action and the need to make monthly decisions on a timely basis, it is critical to develop a planning infrastructure that can provide rapid supply and demand decision support. Further, careful screening is required to assure that all decisions are made at the lowest level possible, in order to assure that a slow, unresponsive planning process does not develop (e.g., decisions that affect global capacity must be made at a global level, while decisions that utilize and affect only regional capacity can be made at the country/region level).

In this section, we have briefly reviewed what has developed into a major hierarchical planning challenge in recent years; namely, integrated global sales and operations planning. Techniques reviewed earlier in this chapter such as hierarchical inventory and forecasting methods play a key role in this planning process. Methodologies such as an ABC approach to inventory management must be applied at both the lower and higher levels of the S&OP process (*i.e.*, at both the country/region level and the global level). Additionally, techniques explored in earlier chapters such as hierarchical production and distribution planning also clearly can play a major role in supporting a firm's global S&OP process.

6.6 Summary

This chapter has explored just a few of the many hierarchical dimensions of inventory management and forecasting. Even these limited examples, however, illustrate the integral role that inventory planning methods (and management) and forecasting play in hierarchical production and distribution planning. Further, a well-conceived hierarchical approach towards these two disciplines represents a prerequisite for any firm desiring to implement and maintain an effective network planning methodology.

Finally, it is important to note that hierarchical approaches are also required in many areas related to sales forecasting such as "sales reporting". For example, monitoring the effectiveness of complex national and international sales force structures requires that a firm organize its sales reporting in an insightful, hierarchical fashion. The hierarchy of reports must highlight both major national and international trends, as well as provide insights to regional and local trends and the effectiveness of "sales force activities".

In the next chapter, we will consider a hierarchical performance measurement approach for monitoring all major aspects of the supply chain including inventory management and sales forecasting.

Chapter 7

Supply Chain And Firm-Wide Performance Measures

Chapters one through six explored hierarchical planning methodologies and algorithms for managing large-scale manufacturing and distribution networks. Key functional areas examined included production planning and scheduling, distribution, transportation, warehousing, inventory management and forecasting. Throughout our discussion, one consistent theme that has prevailed is the need for a firm to maintain an integrated planning and management approach spanning across all functional areas and planning horizons found on supply chains.

Performance measurement systems constitute yet another important planning and management decision support tool that represent a prerequisite to assure effective supply chain operations. In this chapter, we will consider several illustrative performance measurement methodologies that provide a hierarchical, multi-dimensional perspective. We begin in section 7.1 by presenting a hierarchical framework for capturing and linking all key performance measures that a firm may develop to monitor its supply chain. This framework differentiates measures both by their individual level in the hierarchy (*i.e.*, operational, tactical or strategic), and by their focus (e.g., internal organizational efficiency or external effectiveness with customers). In section 7.2, we review some sample logistics performance measures to provide additional background on this subject. Section 7.3 then considers a popular measurement framework frequently employed to monitor overall firm performance; namely, the balanced scorecard (Kaplan and Norton, 1992). Supply chain performance represents just one portion of the collective set of measures typically developed to assess general firm performance in the balanced scorecard. However, our review of the balanced scorecard will focus upon how supply chain performance links into overall firm performance. Specifically, in section 7.4, we illustrate how an organizational unit responsible for a major portion of supply chain operations (e.g., distribution) can create a balanced scorecard-based performance measurement system that quantitatively links the organizational unit's performance to overall firm performance. Section 7.5 concludes this chapter with some final summary remarks.

7.1 A Hierarchical Supply Chain Performance Measurement System

This section presents a three level hierarchical performance measurement framework that can link a wide,array of functional areas and performance measures commonly observed across supply chains. Similar to previous hierarchical systems reviewed in Chapters one through six, this framework has strategic, tactical and operational levels. However, these levels have a different connotation than in previously reviewed frameworks. In particular, the dimension of time or planning horizon does not differentiate the strategic, tactical and operational levels. Rather, it is the scale of an operation or activity that a particular performance measure monitors which determines its place in the hierarchy. Also within the hierarchy of the firm, the level of an organizational or functional unit that a measure monitors determines where it (*i.e.*, the measure) falls in the performance measurement hierarchy.

Within each of the three levels of the measurement system, we further differentiate performance measures into two categories: (1) external measures and (2) internal measures. External measures focus upon the *effectiveness* of an activity or function while internal measures evaluate the *efficiency or productivity* of an activity or function. In particular, external measures evaluate the effectiveness of flows and links across a supply chain, while internal measures evaluate the cost or efficiency of a function or organization in producing its outputs and services. Two typical external performance measures are order and line item fill rates on customer orders. When a customer (e.g., a mass merchandiser) places an order on a supplier (e.g., a manufacturer), the order and line item fill calculations measure whether the supplier delivers the total order and the individual line items on time and complete as ordered. These measures do not evaluate the supplier's cost in delivering the order. Thus, if a supplier had to deliver an order by expensive air freight rather than normal surface transportation because of inventory shortages or other problems, the order and line fill calculations would not capture this. If the shipment arrived on time and complete by air transport, the order fill and line fill measures would identify this particular shipment as a successful fulfillment of a customer order (*i.e.*, the manufacturer was "effective" in delivering the order on time and in full to the customer). In contrast, internal performance measures focusing on the efficiency of an activity would evaluate this order fill example as an "ineffective" effort. "Distribution cost per case" and "freight cost per lb." represent two common internal measures, and this air freight delivery would increase the firm's average distribution cost per case and freight cost per lb. (*i.e.*, it would negatively impact these two performance measures).

Figure 7.1 depicts an integrated hierarchical supply chain performance measurement framework. As previously described, the framework contains three levels and within each level (e.g., strategic) it has both external and internal measures.[1] Note that the framework in Figure 7.1 spans an entire supply chain from the echelon of the raw material and component suppliers, to the echelons of the plants and distribution centers of a manufacturing firm, to the echelon(s) of the locations of the customers of a manufacturer, and finally to the retail consumer. A real world supply chain may have more or less echelons than shown in Figure 7.1. For example, the locations of a manufacturer's customers may span several echelons including receiving warehouse locations and actual retail outlets in the case of a large mass merchandiser. One can adapt the performance measurement framework in Figure 7.1 to as many or as few echelons as may exist on a supply chain. For illustrative purposes, we will now consider this framework from the more limited

[1] Figures 7.2 and 7.3 will provide examples of the measures themselves shortly.

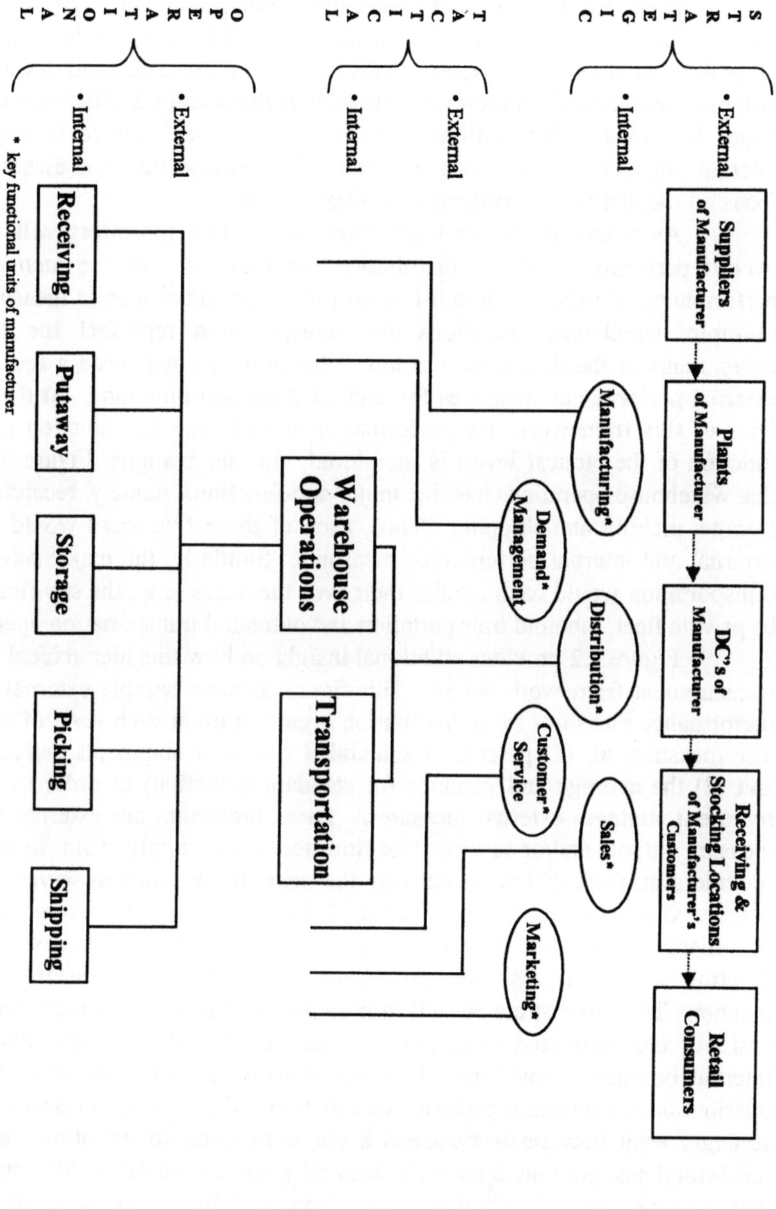

Figure 7.1 Integrated Supply Chain Performance Measurement Framework: Hierarchical Illustration

viewpoint of a single firm (e.g., a manufacturer that wishes to apply this approach to its two echelon network of plants and distribution centers).

Note that Figure 7.1 displays the major functional areas of a firm at the strategic level of this performance hierarchy. (A hierarchy could naturally contain other functional areas not shown in this figure - e.g., research and development.) In order to measure and monitor both the effectiveness and the efficiency of each of its major functions, a firm will want to have a few key high level measures, both internal and external, for each function. To describe this framework, we briefly focus on the distribution organization in particular.

As noted, at the strategic level, a few key indicators will measure the overall performance of the distribution organization. At the *tactical level*, the performance of each major sub-function of a functional area is measured. In this example, warehouse operations and transportation represent the two primary components of the distribution group. Therefore, we will need a few internal and external performance measures for each of these sub-functions. At the *operational level* of this framework, the performance of each key sub-function (of each sub-function of the tactical level) is monitored. In this example, Figure 7.1 illustrates that warehouse operations has five major sub-functions; namely, receiving, putaway, storage, picking and shipping. Thus, each of these five areas would require both external and internal performance measures. Similarly, the major sub-functions of transportation would each require their own measures (e.g., the sub-functions might be private fleet, inbound transportation and outbound transportation operations).

Figure 7.2 provides additional insight on how this hierarchical performance measurement framework works. This figure displays sample external and internal performance measures for a distribution organization at each level of the hierarchy. The measures of: (1) percent of scheduled customer shipments delivered on-time, and (2) the average and variance (or standard deviation) of order cycle lead-time represent strategic external measures. These indicators are external because they measure outputs and/or services that flow across the supply chain. In this case, they provide a measure of how effectively the distribution function serves its customers at the next echelon of the supply chain. These measures are classified as strategic in this hierarchy because they evaluate one of the key missions of a distribution function; namely, to distribute product and/or services to its customers in a timely manner. The other strategic indicator shown in Figure 7.2 is the total distribution cost per unit delivered (e.g., per case delivered). We classify this measure as internal because it has "inward" focus; namely, it evaluates how efficiently the distribution organization performs its function. This internal measure falls into the strategic level because it evaluates a major function (distribution) at a summary level (total cost per unit delivered). Naturally, there exist many different cost factors that comprise total distribution costs. However, in this example, the distribution cost per case delivered provides a high level view of overall distribution cost and effectiveness.

Level:	Type:	Measure:
		Distribution
Strategic	External	■ Percent of scheduled customer shipments delivered on-time
		■ Order cycle lead time: from release to distribution to customer delivery - time in days - variability
	Internal	■ Total distribution cost per unit delivered
		Warehouse Operations
Tactical	External	■ Percent of lines/orders picked correctly
		■ Percent of orders picked on scheduled day
	Internal	■ Total warehouse costs per unit of throughput
		Receiving
Operational	External	■ Percent of cases/lines received correctly
	Internal	■ Total receiving costs per unit

Figure 7.2 Illustrative Hierarchical Performance Measures For Distribution Function

The performance measure examples shown in Figure 7.2 for *warehouse* operations (at the tactical level) and warehouse *receiving* operations (at the operation level) follow a similar theme. The percent of lines or orders picked correctly and the percent of orders picked on the scheduled day represent external measures because they evaluate the impact of warehouse operations across the supply chain. Specifically, when a warehouse picks an order correctly, it contributes to the ultimate successful delivery of products to a customer that has placed an order. In contrast, an incorrect pick will result in an unsuccessful delivery to a customer. Similarly, when a warehouse picks an order on-time (*i.e.*, on the scheduled day), this contributes towards a successful on-time delivery of products to a customer. The third tactical measure shown in Figure 7.2, total warehouse costs per unit of throughput, represents an internal measure. In this case, it offers a summary view of the cost (and efficiency) of the total warehouse operation. Again, the cost of warehouse operations has an "inward" focus as it primarily represents a concern of the distribution/warehouse function (and the manufacturer) rather than the "external" customer at another echelon of the supply chain who receives product from the warehouse (and the manufacturer).[2]

At the operational level, the warehouse receiving function uses the percent of cases (or lines) received correctly (*i.e.*, accurately) as an external performance measure. We categorize this measure as external because the accuracy with which this function receives inbound shipments will impact the next stage of the supply chain. For example, suppose that the receiving area miscodes an inbound receipt as product A, when if fact it received product B on a delivery. If this error remains undetected, the shipment will then be put into inventory classified as product B and at some future point, perhaps could even be picked and delivered to a customer who ordered product A. Regardless of whether this error is caught before the product makes it all the way to the customer, the incorrect receipt impacts downstream stages of the supply chain; and therefore, represents an external measure. In contrast, total receiving cost per unit has an internal orientation and will be of most immediate concern to receiving and warehouse personnel.

Benefits Of A Hierarchical Performance Measurement Methodology

Employing a hierarchical performance measurement framework offers a number of important benefits. First of all, this approach provides a unified framework for aggregating performance measures across a company. In any firm one generally finds a wide-array of performance measures in use in different functional areas. Often even within a particular functional area (e.g., distribution), one finds different measures being used in different regions or countries. By implementing a hierarchical measurement approach, a firm positions itself to organize its existing and future key performance measurements into a structure that leads to a relatively few, high level, strategic measures (e.g., between 10 and 25) that monitor overall

---

[2] It is important to note that although this internal measure is primarily of interest to warehouse operations (and the manufacturer), in the long run one can argue that it is of interest to all participants in a supply chain. This is because the end customer (e.g., a mass merchandiser/retailer) and the end consumer (e.g., the consumer who purchases from the retailer) desire the lowest purchase cost possible. The efficiency of all upstream participants in a supply chain (e.g. a manufacturer and all its functional areas) ultimately has a significant impact on the purchase price of the end customer and end consumer.

firm performance. At the same time, this structure facilitates having many other performance measures that monitor smaller components of a firm's operation, yet which align with overall firm objectives. In particular, a hierarchical measurement system allows both large and small functional areas within a firm to:

1. develop and maintain their own measures, and
2. contribute to and be part of an overall measurement system

This hierarchical approach thus helps to keep measures both "simple" and "meaningful" because each function and sub-function at each level can focus on a few key performance measures. For example, colleagues in the receiving operation of a warehouse can focus on the few measures that they develop to monitor their activity. These measures will be meaningful to the receiving colleagues because they (the colleagues) will strongly influence how good or bad a performance these measures actually depict. At the same time, because these operational level measures are components and contributors to an overall measurement system, colleagues in an operational function such as receiving can see how their activity fits into the overall operations of the firm. Thus, if properly implemented, this hierarchical approach can also make the higher level strategic measures of a firm (e.g., customer service order fill rates) meaningful to all colleagues of the firm.

A hierarchical performance measurement system can also contribute towards aligning the collective activities of a firm to meet a desired mission and set of objectives. For example, if a firm has a comprehensive measurement system in place that covers most or all of its major functional areas and activities, managers can view the system in its entirety to identify any potential misaligned activities or objectives. Specifically, if the desired results for performance measures in one area (e.g., manufacturing) contradict what would represent desired results for performance measures in a second area (e.g., sales); this generally would represent a tip-off to explore further how well these two functional areas are working together to meet overall firm objectives. If the "known" performance measures of different functional areas are in conflict with each other, then it is quite likely that the less visible individual performance incentives of managers in different functional areas may also be in conflict.[3]

Finally, in order for a firm to employ this performance measurement framework most effectively, similar to all hierarchical systems, it requires that its users create and utilize feedback loops between the different levels, and across different functions. A simple example of feedback within a fictitious distribution organization will serve to illustrate this concept. Assume that distribution has responsibility for the order cycle process from the receipt of a customer's order until delivery of that order to the customer's receiving location. Assume further that distribution has a strategic level performance goal of four days total order cycle time (from receipt of a customer order until delivery of that order). Now however,

[3] In reality, one invariably finds seemingly conflicting goals between different functional areas within a firm. What is important is to understand these potential conflicts and to make sure that planned firm actions recognize these conflicts and in the end develop synchronized plans. A highly visible and utilized hierarchical performance measurement system facilitates this process.

assume that at the operational and tactical levels, the following performance goals exist:

- order receipt to release to warehouse and transportation scheduling -1 day
- warehouse and transportation scheduling -1 day
- warehouse picking and shipping -1 day
- transportation transit time from DC to customer's location -2 days

The above example suggests that that when all the operational and tactical sub-functions of distribution meet their order cycle performance measurement goals, distribution can achieve an order cycle time of five days. However, we previously observed that distribution has an overall performance objective of four days for total order cycle time. Feedback of goals from the operational and tactical levels is required in this example to highlight that the overall strategic performance objective of distribution is not achievable based on the current performance standards and goals at lower levels of the organization. Although this represents an overly obvious example, it illustrates the sort of feedback that is required to assure that lower level goals and performance align with overall organizational goals and performance.

7.2 Logistics And Supply Chain Performance Measures

This section present a selection of common performance measures employed to monitor and evaluate the supply chains of large-scale networks. It provides additional background on the types of metrics that populate hierarchical performance measurement systems.

An extensive literature on logistics and supply chain performance measurements exists today, and this literature continues to expand at a rapid pace. (The interested reader is referred to the following articles for a brief sample of this literature: van Amstel and D'hert, 1996; Fawcett and Cooper, 1998; Johnson and Davis, 1998; and Mentzer and Konrad, 1991). Generally, both authors who contribute to the literature, and firms in practice, group logistics performance measures into major categories. Further, in both academic and industrial settings, one can find a wide variety of classification schemes for performance measures. Two typical examples of classification schemes are those of Fawcett and Cooper (1998), and Mentzer and Konrad (1991). Figure 7.3 depicts their respective schemes.

Mentzer and Konrad	**Fawcett and Cooper**[1]
• Transportation	• Asset Management
• Warehousing	• Cost
• Inventory Control	• Customer Service
• Order Processing	• Productivity
• Logistics Administration	• Logistics Quality

[1] This set of categories was also proposed by Bowersox and Closs, 1996, and by Bowersox et. al. 1989

Figure 7.3 Common Logistics Performance Measurement Categories

The first set of categories (Mentzer and Konrad) groups logistics performance measures by functional area, while the second set of categories (Fawcett and Cooper) is a generic set that can be applied across logistics functions. Thus, one can apply the second group to each individual logistics function. Fawcett and Cooper illustrate this by applying their categories to the transportation function as follows:

- *Asset management*: Capacity utilization, return on investment
- *Cost:* Cost per hundredweight, transportation cost per unit
- *Customer Service*: Average transit time, transit time variability
- *Productivity*: Orders delivered per vehicle, full versus partial loads
- *Quality*: Damage in transit, documentation accuracy

Thus, using the scheme of Fawcett and Cooper, one could apply these generic categories to warehousing, inventory management and so on.

In developing groups to categorize individual performance measures, one factor to consider is that the demand for information and measurements on supply chain performance has increased dramatically in recent years. Additionally, retail customers such as mass merchandisers, wholesale clubs, etc. have become extremely demanding of their suppliers in terms of quality of overall logistics performance required (e.g., on-time delivery, line and order fill rates, and so on). Similarly, suppliers have become much more precise in measuring their costs to serve individual customers, and in incorporating these costs into "bundled" product price and service offerings.[4] The net result of this increased availability and focus on information, and the increased logistics services requirements of partners across the supply chain; is that it has become more difficult to define one standard set of supply chain/logistics measurement categories that works well for most firms. To develop a set of supply chain performance categories generally requires more time and effort (and customization) than was necessary 10 to 20 years ago. Nevertheless, defining an appropriate set of categories that meets the needs of a particular firm or set of firms represents a critical step towards establishing an effective performance management system.

Individual Performance Measures

There exist a wide array of supply chain performance measures employed in practice, and the relative importance of specific measures varies by firm. Table 7.1 offers a sample of some performance measures commonly used across many industries. This list was developed by Fawcett and Cooper (1998) as part of their research on trends in the availability of logistics information. In this research, the authors surveyed over 100 major companies in North America, Europe and the

[4] The 'Streamlined Logistics" program of The Proctor and Gamble Company is one of the best known examples in US industry of a program that "bundles" goods and services together for a specific price. Importantly, this program and others similar to it, also make very specific requirements of customers. For example, to obtain the lowest price available, customers typically must meet requirements such as they must unload the suppliers' trucks within several hours once the trucks arrive, orders must be placed electronically and must provide detailed order information (e.g., expected prices), etc. These "bundled requirements" are in addition to traditional order quantity minimums required to obtain best price.

Pacific Basin and queried each firm as to the availability of specific performance measures at these firms. One question that the authors addressed was how the availability of information had improved between 1989 and 1994. Table 7.1 describes the results of the research on this question, in addition to illustrating the types of individual performance measures that populate a hierarchical supply chain performance measurement system. Note that a number of the measures in Table 7.1 could appear at multiple levels of a hierarchy and also could appear in multiple logistics functions and sub-functions (e.g., order cycle is typically monitored in total, and by the major components that comprise total order cycle time).

Table 7.1 Illustrative Logistics Performance Measurements

	Information Availability	
MEASURES	1994	1989
Asset Management		
Inventory levels	97.3%	89.9%
Obsolete inventory	95.9	86.4
Inventory turns	94.6	90.2
Return on net assets	88.9	80.7
Inventory classification	87.3	NI
Return on investment	86.3	84.8
Cost		
Comparison of actual versus budget	100.0	92.4
Outbound freight	97.4	94.3
Cost as a percent of sales	94.6	91.5
Total cost (logistics)	93.5	84.1
Administrative	93.3	85.1
Direct labor	92.0	85.1
Cost trend analysis	89.6	73.7
Cost per unit	86.5	82.0
Cost of damage	81.6	66.8
Warehouse order processing	81.6	62.4
Inventory carrying cost	79.2	76.8
Inbound freight	76.7	87.3
Cost of returned goods	71.1	68.5
Direct product profitability	54.1	58.1
Customer or segment profitability	47.9	NI
Cost of service failure	41.3	NI
Cost of backorder	31.0	NI
Productivity		
Warehouse labor productivity	86.8	NI
Comparison to historical standard	85.1	83.7
Goal programs	83.1	76.2
Units shipped per employee	71.6	70.3
Productivity index	71.5	59.2
Units per labor dollars	63.5	67.8
Order entry productivity	56.5	NI
Equipment downtime	56.2	NI
Orders per sales representative	52.3	59.5
Transportation labor productivity	62.2	NI
Customer service		
On-time delivery	88.0	74.8

Table 7.1 Illustrative Logistics Performance Measurements (continued)

Stockouts	88.0	79.9
Shipping errors	85.3	80.5
Fill rate	85.1	79.4
Delivery consistency	83.8	NI
Cycle time	82.7	61.7
Backorders	74.3	76.4
Complete orders	73.6	NI
Customer complaints	73.0	81.3
Overall customer satisfaction	56.8	NI
Overall reliability	51.4	NI
Response time to inquiries	37.5	NI
Sales force complaints	36.8	76.5
Response accuracy	26.0	NI
Logistics quality		
Number of customer returns	93.2	78.0
Picking/shipping accuracy	83.1	NI
Number of credit claims	82.9	75.7
Damage frequency	74.0	58.6
Document/invoicing accuracy	61.8	NI
Order entry accuracy	61.6	NI
Information availability	51.4	NI
Information accuracy	38.9	NI

We close this section with two final comments on performance measures. First, it is important to recognize that logistics performance should be measured across all echelons of a supply chain. This represents a major task to complete and naturally a first step in this process for any firm consists of developing an effective measurement system to monitor its own performance. Conceptually, however, this measurement process should extend across a total supply chain often consisting of several firms. Figure 7.4 illustrates how specific performance indicators can be measured at various echelons of a supply chain. In this example, we assume that the firms or participants on this supply chain are: (1) the raw material suppliers to a manufacturer, (2) the manufacturer, (3) the retail customers of the manufacturer (e.g., mass merchandisers), and (4) the end consumers who purchase the manufacturer's products at the retailers' stores. Note that the manufacturer has two echelons in this figure (its plants and its distribution centers), and the retailers also have two echelons (their distribution centers and their stores). For simplicity, we illustrate this concept for only a few customer service performance measures.

Finally, we should emphasize that the complexity and potential dimensions of a supply chain performance measurement system have increased rapidly in recent years as networks and companies have become truly global. Just as the S&OP process reviewed in Chapter 6 has taken on heightened complexity in a global environment, so too has supply chain performance measurement. This creates an even greater imperative to establish a well-defined hierarchical framework to house a firm's performance measurements. A well-constructed framework facilitates the effective capture and organization of performance measurements covering ever-expanding geographic and other supply chain dimensions.

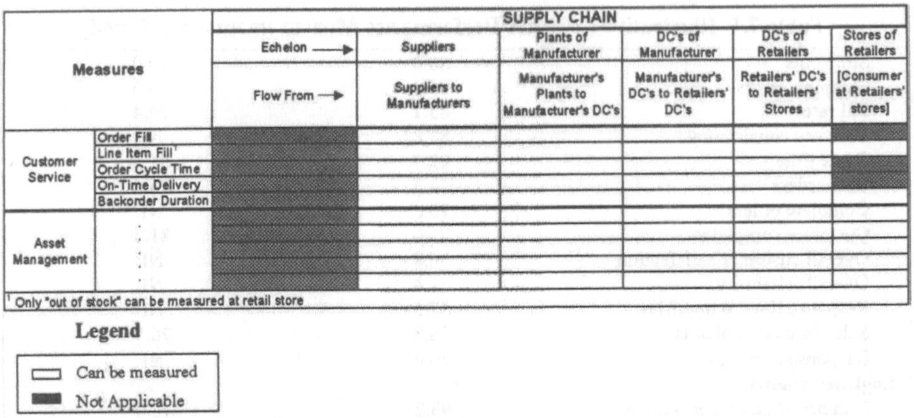

Figure 7.4 Measuring Logistics Performance Across A Supply Chain (Illustration of Echelons And Flows Where Measures Apply)

7.3 The Balanced Scorecard: A Firm-Wide Performance Measurement System

A balanced scorecard (Norton and Kaplan, 1992) should monitor a firm's overall performance and its progress in meeting short and long-term objectives. Given the integral role that a manufacturing and distribution network plays in a firm's operations, a firm's performance as measured by a balanced scorecard is heavily influenced by the effectiveness and efficiency of its supply chain. In this section, we will review the balanced scorecard, which is itself a hierarchical performance measurement system. This discussion will facilitate the introduction in section 7.4 of a methodology that links supply chain operations directly to a firm's balanced scorecard.

The Balanced Scorecard

The balanced scorecard developed by Kaplan and Norton has created enormous interest in recent years. It is based on the idea that managers have to evaluate their business from at least four major perspectives: customers, internal business, innovation and learning, and financial. According to Kaplan and Norton, the performance measures developed to monitor these four perspectives should answer the following questions:

(1) How do customers view a firm?
(2) What business processes must the firm improve and exceed at?
(3) Can the firm continue to learn, improve, and thereby create value?, and
(4) How does the firm appear to its shareholders?

The balanced scorecard is intended to provide managers with a streamlined view of most major activities. The ability of the balanced scorecard to provide this view depends upon the construction of a set of performance measures which capture the pulse of a corporation in a few, focused indicators. This is a challenging task.

Figure 7.5 presents an illustrative balanced scorecard with a fairly typical set of performance measures. (Kaplan and Norton, 1993, review alternative balanced scorecards developed by several firms.) Each performance measure addresses a particular goal or capability under one of the four "firm" perspectives. For example, cash flow (Financial) provides an important indication of a firm's ability to sustain itself, particularly in the near term. In contrast, trends in ROE, ROCE, and market share are important barometers of long-term health. Similarly, order fill rates and line fill rates (Customer) provide insights into the operational capabilities of a firm to deliver product consistently. Customer partnerships and satisfaction provide a perspective on how well a firm is positioned for a sustained, prosperous relationship with its customer base. There is no absolute "right" set of performance measurements that every firm should employ. In general, however, the more comprehensive yet concise the set of measures, the greater is the probability that the balanced scorecard will provide an effective monitoring and management tool for a firm. (The reader is referred to Kaplan and Norton, 1992 and 1996, for a detailed review.)

CUSTOMER PERSPECTIVE	
Goals	**Measures**
Consistent, timely supply	Order fill rate Line fill rate
Partnership relationships	Number of customer partnerships
Satisfied customers	Customer surveys

INNOVATION AND LEARNING	
Goals	**Measures**
Compressed, responsive product development	Product development to market cycle time
New product leadership	Percentage of sales from new products
Manufacturing excellence	Manufacturing yield

INTERNAL BUSINESS	
Goals	**Measures**
Operational excellence	Cycle time: manufacturing to customer delivery Cost per normalized unit
Reliable new product introduction	New product actual vs. planned introduction
Satisfied, productive employees	Employee survey

FINANCIAL	
Goals	**Measures**
Sustain business	Cash flow
Grow business	Operating income Sales growth
Long-term success	Market share ROE

Note: The format source is Kaplan and Norton 1992.

Figure 7.5 Illustration Of A Balanced Scorecard

7.4 Integrating The Balanced Scorecard And Supply Chain Performance Measurement

This section describes a methodology for developing a direct quantitative link between the performance of a firm and its major organizational units, and the firm's balanced scorecard. In the example that follows, we will apply this methodology to the distribution unit of a firm for illustrative purposes, given that distribution typically represents a major participant in supply chain activities. A principal reason for developing the quantitative approach to follow stems from the author's empirical observation that with the exception of financial measures, managers often have difficulty relating the balanced scorecard and its performance measures to overall firm performance. This has led to many situations where a firm invests substantial time and resources in developing a balanced scorecard, only to have the scorecard ignored by the organization, or at best become a source of passing interest, heavily underutilized by the firm.

To facilitate the development of a quantitative methodology linking the balanced scorecard to a firm's overall performance, we first turn to a popular hierarchical strategic planning approach known as the "mission, objective and strategy" (MOS) method (see Liberatore, Monahan and Stout, 1992). Figure 7.6 illustrates the basic MOS methodology. Briefly, it employs a three-tier planning hierarchy. The first level - the firm's overall mission - often is expressed as the maximization of stakeholder or shareholder value subject to satisfying some defined set of product and service requirements for a defined customer base and market. In the second stage of this approach, the firm evaluates its internal and external position, and then based on this evaluation, the firm defines objectives for achieving its stated mission. Finally, in the third planning stage, the firm develops strategies designed to meet those objectives posed in the second stage.

One can easily tailor the MOS approach for our purposes to the development of the strategies of the distribution organization. Simply put, the distribution strategy should support the overall objectives and mission of the firm. Thus, the distribution strategy simply represents a subset of the firm's overall set of strategies. This allows us to substitute "distribution strategies" for "strategies" in the MOS framework. We then can construct an overall framework for integrating a firm's balanced scorecard with the mission, objectives and strategies of the firm, as depicted in Figure 7.7.

In this framework, the balanced scorecard's performance measures track how successfully a firm is carrying out its strategies, objectives, and overall mission. It is the report card, from an array of perspectives (financial, customers, and so on) on the firm's progress toward its MOS. Thus, the balanced scorecard essentially plays the role of the thermometer in a firm's MOS. In a sense, the performance measures of a balanced scorecard also represent a quantitative set of "surrogate" measures of whether a firm is achieving its mission. Thus, distribution strategies developed to carry out a firm's MOS will naturally affect performance measures, as Figure 7.7 depicts.

In summary, Figure 7.7 offers a conceptual framework for understanding the role played by a balanced scorecard in the planning and management of a firm's MOS.

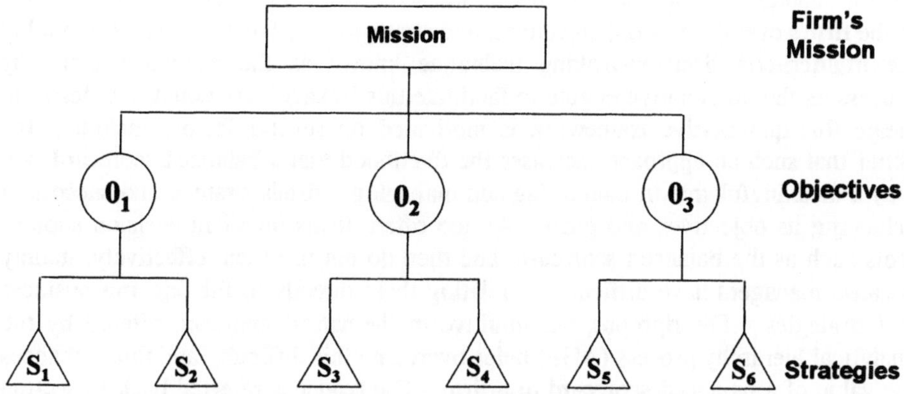

Figure 7.6 The Mission, Objective And Strategy (MOS) Approach To Strategic Planning

EXAMPLE: The Relationship Of The Balanced Scorecard To Distribution Strategies And The Firm's Overall Mission and Objectives

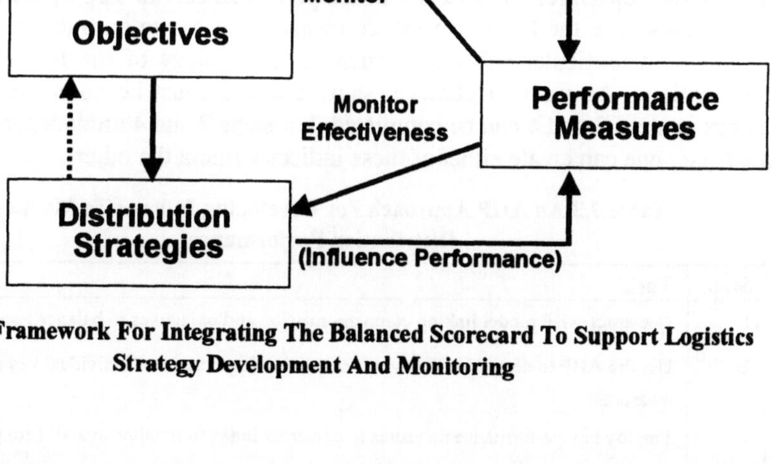

Figure 7.7 Framework For Integrating The Balanced Scorecard To Support Logistics Strategy Development And Monitoring

The Analytic Hierarchy Process Framework For Integrating The Balanced Scorecard And The MOS

Now that we have positioned the balanced scorecard in the MOS framework, we need a quantitative approach for linking the scorecard's key performance measures to the firm's overall mission, objectives, and strategies. For that purpose, we employ the multicriteria decision-making technique known as the analytical hierarchy process as the quantitative engine to facilitate this linkage. As noted, the desire to create this quantitative framework is motivated by several factors including the belief that such an approach increases the likelihood that a balanced scorecard will play a meaningful role in monitoring and managing a firm's strategy and success in achieving its objectives and goals. All too often, firms invest in decision support tools such as the balanced scorecard, and then do not use them effectively, mainly because managers have difficulty in relating them directly to the ongoing business and strategies. The rigorous, yet intuitive mathematical approach offered by the analytical hierarchy process (AHP) helps overcome this difficulty and thus enhances the value of a balanced scorecard to a firm. The reader is referred back to section 4.3 and also Saaty (1996) for a review of the AHP methodology.

The AHP-Based Framework

Table 7.2 details the steps required to link the performance measures of the balanced scorecard to (1) the overall mission and objectives of the firm, and (2) the strategies of a particular unit of the firm (such as distribution) designed to help achieve the firm's objectives and mission. This process begins with the construction of an AHP model that links, as we will discuss shortly, the firm's performance measures directly to its objectives and mission (step 1). After creating this linked hierarchy, we then use the AHP to determine the relative weight or importance of each individual performance measure to the firm's ability to succeed in its objectives and mission (step 2). The relative weights of the performance measures determined in step 2 then allow us to create an index (step 3) which collectively uses the key performance measures to track the firm's performance. In step 4, we show how a firm can also use the key performance measures to create an index to monitor the strategy of a particular sub-group, such as the strategy of the firm's distribution organization. As we will observe, steps 1 and 2 must be completed before the indices in steps 3 and 4 can be completed, but steps 3 and 4 are independent of each other (*i.e.*, one can create either of these indices without the other).

Table 7.2 An AHP Approach For Developing Indexes To Monitor Firm And Distribution Performance

Step	Task
1	Construct AHP model linking company mission and objectives to balanced scorecard
2	Use the AHP to determine "weights" or relative importance of individual key performance measures
3	Employ key performance measures to construct index to monitor overall firm performance
4	Use "distribution-oriented" performance measures to construct index to monitor distribution strategy

Step 1: Construct AHP Linking Company Mission and Objectives to the Balanced Scorecard

We employ a four-level hierarchy to link the performance measures of the balanced scorecard to the firm's mission and objectives. These four levels are from top to bottom:

1. The firm's mission
2. The firm's objectives
3. The four components of the balanced scorecard
4. The key performance measures of the balanced scorecard

Figure 7.8 illustrates this hierarchy. The top tier displays the goal of this AHP design. An AHP model must have an explicit goal, and in our case it is "to determine the relative importance of the individual performance measures to the firm's mission and objectives." Thus, Figure 7.8 represents the AHP model we will employ. Implicitly, however, the top tier also represents the company's mission. For example, assume that the firm has the following mission:

Maximize shareholders' and all stakeholders' value by being a leader in the manufacture and distribution of XYZ products and services.

To visualize the firm's four-tier hierarchy, one can simply substitute a box containing the firm's mission for the box labeled "Goal" in Figure 7.8. Now let us explore how we will use this hierarchy.

As reviewed in section 4.3, AHP models are based upon pairwise comparisons and relationships between different levels of the AHP hierarchy. Consider the hierarchy in Figure 7.8, and assume that the firm has (1) developed a balanced scorecard, which includes all the key performance measures shown in this figure, and (2) determined its key objectives (as shown).

We would begin our modeling process by first asking management to evaluate the relative importance of each of these critical objectives in carrying out the firm's mission. For example, with respect to that mission, management would be asked how much more important is increasing growth and market share relative to increasing profitability. The same question would then be asked to compare the relative importance of growth and market share to improving customer satisfaction. This process would continue until all six possible pairwise comparisons at the "objectives" level have been completed.

Management next would have to address the relative importance of each of the four components of the balanced scorecard with respect to meeting the firm's four objectives. For example, with respect to the firm's objectives for growth and market share, how much more important is the firm's financial performance relative to its performance in meeting customer expectations? Again, comparisons would be conducted for all potential unique pairs at this third level of the hierarchy (six comparisons for each of the four objectives, or twenty four in total). Finally, similar

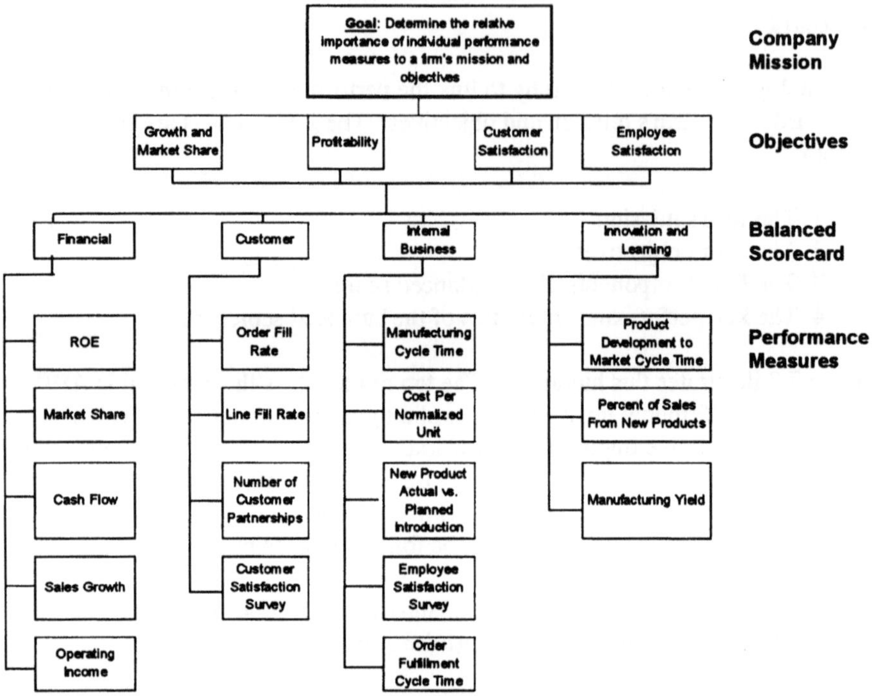

Figure 7.8 AHP-Balanced Scorecard Framework

pairwise comparisons would occur at the fourth and lowest level of the hierarchy. For example, with respect to measuring the financial performance of the firm, management would be asked to address the relative importance of sales growth versus operating income. Similar questions would be asked about the relative importance of all performance measures under the financial, customer, internal business, and innovation and learning categories. Using the synthesis process of the AHP, the relative assessments made across the hierarchy are combined, employing a weighted average approach, to determine the relative effect of each measure in meeting the firm's mission.

A number of different approaches are available to facilitate the development of the relative weights required in the AHP hierarchy. Briefly, one of the first decisions is to determine who in a firm will contribute to the "weighting" development process. Clearly, there is no single correct approach. Based on previous experience, however, we suggest that a mix of general senior managers and functional senior managers (e.g., manufacturing, logistics, and sales and marketing) represents a good group to perform the "criteria weighting" task. A well-balanced combination of senior managers should have a good overall perspective on what factors (*i.e.*, criteria in AHP lexicon) most influence a company's ability to meet its objectives and carry out its mission. Once the group is chosen, a voting method for deciding the actual AHP-balanced scorecard weights must be selected. There are several possibilities. For example, suppose that the process begins with each group member first independently developing a complete set of pairwise comparisons for the criteria. A method is required to combine these judgment sets in such a way that a consensus is obtained. Two acceptable approaches are as follows.

One way to combine these judgments is for the group to resolve any differences by seeking a consensus through discussion. A second way is to compute the geometric mean of each of the judgments. For example, suppose there are five group members, and each has provided a judgment comparing the relative importance of the growth and market share objective to the profit objective (in the second level in the AHP hierarchy) with respect to achieving the firm's mission (the first level of the hierarchy). If 3, 2, 4, 3, and 4 are the five judgments, then their geometric mean is the fifth root of their product ([(3) (2) (4) (3) (4)] 1/5), which equals 3.1. The value of 3.1 is entered into the pairwise comparison matrix in the appropriate cell. This process is repeated for all judgments as needed.

Step 2: Use the AHP to Determine the "Weights" or Relative Importance of the Individual Key Performance Measures

The pairwise comparisons described in step 1 can be entered and processed with a user-friendly AHP software package such as Expert Choice for Windows 9.0 (Foreman et. al., 1990). Expert Choice enables the user to quickly and easily structure hierarchies, enter all necessary judgments, and automatically compute the alternative and criteria weights. After structuring the hierarchy, Expert Choice offers several modes for entering judgments, including verbal and numerical approaches. This software package automatically computes the weights and a measure of the consistency of judgments. It also provides the user with assistance in identifying the most inconsistent judgments. In addition, Expert Choice offers a full range of sensitivity analysis capabilities.

Table 7.3 displays the illustrative relative weights of the key performance measures as calculated by the AHP. For example, cash flow (P3) has a relative weight or importance of 6% (among all the performance measures), line fill rate (P7) has a weight of 7%, employee satisfaction (P13) has a weight of 4%, and so on. Note that the relative weights of the full set sum to 100%. Again, these weights relate the relative importance of each performance measure to the firm's efforts to meet its overall mission. Thus, using the AHP, we have now quantitatively linked each performance measure to the overall firm mission.

Table 7.3: Relative Weights Of Performance Measures Determined By AHP

Performance Measure		Weight
P1.	ROE (%)	7%
P2.	Market Share (%)	7%
P3.	Cash Flow ($000)	6%
P4.	Sales Growth (%)	6%
P5.	Operating Income ($000)	7%
P6.	Order Fill Rate (%)	6%
P7.	Line Fill Rate (%)	7%
P8.	Number of Customer Partnerships (#)	3%
P9.	Customer Satisfaction Survey (Rating)	7%
P10.	Manufacturing Cycle Time (Days)	7%
P11.	Cost per Normalized Unit ($/Unit)	7%
P12.	New Product Actual vs. Planned Introduction (Days Past Plan)	3%
P13.	Employee Satisfaction Survey (Rating)	4%
P14.	Order Fulfillment Cycle Time (Days)	7%
P15.	Product Development to Market Cycle Time (Months)	7%
P16.	Percent of Sales from New Products (%)	3%
P17.	Manufacturing Yield (%)	6%
		TOTAL 100%

Step 3: Employ Key Performance Measures to Construct an Index to Monitor Overall Firm Performance

The relative weights for each performance measure constructed in step 2 facilitate the development of an index to track the firm's progress in carrying out its mission. The concept underlying this index, which builds upon a method suggested by Liberatore, Monahan, and Tymon, (1997) is as follows. By means of the AHP model, we now know the relative importance of each performance measure to the success of the firm's mission. We can also track the firm's progress against each performance measure from period to period (e.g., quarter to quarter). For example, how does cash flow (P3) in one period compare to cash flow in the previous period? Because we know the relative importance of each performance measure to the firm's overall mission, we can assess the relative effect of the change in a performance measure from one period to the next on the firm's overall performance. By individually tracking each performance measure from period to period, and then collectively quantifying the relative effect of the change in each performance measure, we can construct an index to monitor the firm's progress against its mission.

The method used to develop an index to track overall firm performance is essentially identical to that which we will use to establish an index to monitor the firm's distribution strategy (step 4). The only difference is that in establishing our distribution index we use only the subset of performance measures directly related to the firm's distribution performance. (To monitor the firm's overall performance, we use all performance measures of the balanced scorecard.) Because our focus here is the supply chain, we now proceed directly to the construction of an illustrative distribution index. (Appendix 7.4A demonstrates how to develop an overall index to monitor firm performance.)

Step 4: Use Distribution-Oriented Performance Measures to Construct an Index to Monitor the Firm's Distribution Strategy

To create an index for tracking the firm's distribution strategy, we begin by selecting the subset of performance measures most pertinent to distribution. The first column of Table 7.4 displays an illustrative subset. We chose eight of the seventeen total performance measures for this "distribution" subset. (One could theoretically devise a more rigorous methodology to select the subset of "distribution" oriented performance measures. For example, one could even construct a second AHP model to select this subset. For simplicity, we choose the subset shown.) It is important to note that in step 4 a firm may choose to have a "functional subgroup" select the subset of performance measures pertinent to its functional area. For example, marketing, sales and distribution employees might choose the performance measures to include in the distribution index. Presumably, such a subgroup would have the best perspective on which factors have the strongest influence on how well a distribution strategy supports a company's objectives and mission. Note that this subgroup could not change the weights previously determined by the first group which constructed the overall AHP. The subgroup would simply select the criteria pertinent to its functional area.

For reasons which we will clarify shortly, the next step consists of "renormalizing" the weights of the logistics measures subset to total 100%. Table 7.4 illustrates that we simply multiply the overall weight of each of the eight measures by the ratio of 100% to the subset's overall weight (50%) to renormalize the logistics subset. We can now build a distribution index.

Table 7.4 Relative Weights Of Performance Measures For Distribution Index

Performance Measure		Overall Weight	Renormalized "Distribution Only" Weight
P4.	Sales Growth (%)	6%	12%
P5.	Operating Income ($000)	7%	14%
P6.	Order Fill Rate (%)	6%	12%
P7.	Line Fill Rate (%)	7%	14%
P9.	Customer Satisfaction Survey (Rating)	7%	14%
P11.	Cost per Normalized Unit ($/Unit)	7%	14%
P14.	Order Fulfillment Cycle Time (Days)	7%	14%
P16.	Percent of Sales From New Products (%)	3%	6%
		50%	100%

The first step in constructing an index is to establish a baseline value or score for the index in an initial or "base" period (such as the first quarter of 2000).

The baseline value provides an index score against which we can then measure the value of the score of subsequent periods. Table 7.5 illustrates the baseline distribution index.

Table 7.5 Baseline Distribution Index

Performance Measure	Baseline Value[a]	Initial Points	Weighted Points
P4. Sales Growth (%)	10.0	100	12[b]
P5. Op. Income ($000)	66.0	100	14
P6. Order Fill Rate (%)	84.0	100	12
P7. Line Fill Rate (%)	96.0	100	14
P9. Cust. Sat. Survey (Rating)	90.0	100	14
P11. Cost Per Norm. Unit ($/Unit)	1.00	100	14
P14. Order Cycle Time (Days)	10.0	100	14
P16. Percent Sales New Prod. (%)	20.0	100	6
BASELINE INDEX			100.0

[a] Value in initial period (quarter) of index
[b] Sample Calculation: 12 = .12 x 100, that is, weighted points = renormalized channels only weight (Table 7.4) x 100

The actual value of a performance measure in the base period represents the baseline value for each individual performance measure. Table 7.5 indicates that, for distribution, the baseline value for sales growth (P4) is 10%, for operating income (P5) it is $66 million, for order fill rate (P6) it is 84%, and so on. To initiate the baseline index, each performance measure receives 100 initial points. Then, as Table 7.5 shows, we multiply the initial points for each performance measure by the renormalized "distribution only" weight of each measure. This creates an initial weighted point value for each measure equal to its renormalized weight. Note that, by construction, the sum of the weighted points for the renormalized distribution subset of measures totals to 100%. This completes the construction of the baseline index for "distribution."

With the baseline index created, we can now generate the actual distribution index. This will serve to track distribution performance over time. Table 7.6 displays an index for the second quarter of 2000. To illustrate how this tracks performance, we review a sample calculation. As shown under current quarter value, sales (P4) increased by 11% during the second period. We calculate the "points" of P4 in 2000.2 by dividing its current value by its baseline value and multiplying by its initial points ([11% / 10%] x 100 = 110). Next, we derive the weighted points of P4 for this period by multiplying the renormalized relative weight of this measure by its points (.12 x 110 = 13.2). After deriving the weighted points for each distribution performance measure, we then sum the weighted points of all measures to determine the index value for the particular quarter. Table 7.6 indicates a distribution index value of 102.8 for 2000 Quarter 2.

Table 7.6 Distribution Index

Performance Measure	Current [a] Quarter Value	Points	Weighted Points
P4. Sales Growth (%)	11.0	110.0[b]	13.2[c]
P5. Op. Income ($000)	70.0	106.1	14.8
P6. Order Fill Rate (%)	85.0	101.2	12.1
P7. Line Fill Rate (%)	97.0	101.0	14.1
P9. Cust. Sat. Survey (Rating)	90.0	100.0	14.0
P11. Cost Per Norm. Unit ($/Unit)	0.98	102.0	14.3
P14. Order Cycle Time (Days)	10.0	100.0	14.0
P16. Percent Sales New Prod. (%)	21.0	105.0	6.3
BASELINE INDEX			102.8

[a] Value in initial period (quarter) of index

[b] Sample Calculation: 110.0 = (11.0/10.0) x 100: *i.e.*, Points = current quarter value/baseline value (Table 7.5) x 100

[c] Sample Calculation: 13.2 = .12 x 110, *i.e.*, Weighted Points = Renormalized channels only weight (Table 7.4) x points

The distribution index offers an important measure by which to track the relative contribution of the firm's distribution strategy to the firm's objectives and mission. In this example, the index value of 102.8 indicates that the firm has progressed from the baseline period (when the index equaled 100.0) in accomplishing its objectives and mission. Over time, the change and rate of change in such an index provides a quantitative perspective on overall progress from period to period.

Finally, we should point out that, for simplicity, we have assumed a linear scale in constructing this illustrative index. That is, a 10% increase in the value of a performance measure from one period to the next will result in a 10% increase in the value of that measure's contribution to the overall index. In certain situations, a linear scale may not be appropriate. Appendix 7.4B provides examples of the limitations of a linear scale and offers an approach for handling more complex situations not well suited to a linear scale.

This completes the construction of the illustrative distribution index. We conclude with the observation that this same approach can be used to develop an index (or indices) for a number of different functional areas or processes within a firm. For example, one could establish an index to track manufacturing operations, marketing activities, research and development, and so on. In each case, one would begin by determining the subset of balanced scorecard performance measures most germane to the function or process. Then one would create an index for the particular function or process in the identical manner used here to generate a distribution index.

Appendix 7.4A Constructing An Index To Monitor Overall Firm Performance

We use the same basic method to create an overall firm performance measure index as we employed to generate the distribution index. The relative weights of each performance measure (refer to Table 7.3) form the starting point to build the index.

Because this index will track the overall performance of the firm, we include all seventeen of the firm's performance measures. In addition, because this index includes all the firm's performance measures, we do not need to "renormalize" the measures' weights (as was required for the distribution index). Thus, the relative weights displayed in Table 7.2 represent the weights used in the index for each measure.

Table 7.7 displays the baseline for the firm's overall performance measure index. The values of each performance measure in the initial quarter provide the baseline values for the index (e.g., a 10% ROE, a 30% market share, and so on). Following the same method used to develop the distribution index, each performance measure initially receives 100 points. The relative weight of the performance measure multiplied by the initial point value yields the baseline weighted point value, shown in the last column of Table 7.7. Again by construction, the baseline index totals to 100.0.

Table 7.7 Baseline Firm Performance Measure Index

Performance Measure		Base Value[a]	Initial Points	Weighted Points
P1.	ROE (%)	10.0	100	7[b]
P2.	Market Share (%)	30.0	100	7
P3.	Cash Flow ($000)	100,000	100	6
P4.	Sales Growth (%)	10.0	100	6
P5.	Operating Income ($000)	66,000	100	7
P6.	Order Fill Rate (%)	84.0	100	6
P7.	Line Fill Rate (%)	96.0	100	7
P8.	Number of Customer Partnerships (#)	30	100	3
P9.	Customer Satisfaction Survey (Rating)	90.0	100	7
P10.	Manufacturing Cycle Time (Days)	10.0	100	7
P11.	Cost per Normalized Unit ($/Unit)	1.00	100	7
P12.	New Product Actual vs. Planned Introduction (Days Past Plan)	40	100	3
P13.	Employee Satisfaction Survey (Rating)	80.0	100	4
P14.	Order Fulfillment Cycle Time (Days)	10.0	100	7
P15.	Product Development to Market Cycle Time (Months)	30.0	100	7
P16.	Percent of Sales from New Products (%)	20.0	100	3
P17.	Manufacturing Yield (%)	95.0	100	6
BASELINE INDEX				100.0

[a] Value in initial period (quarter) of index
[b] Sample Calculation: 7 = .07 x 100; that is, weighted points = overall weights (Table 7.3) x 100

Table 7.8 illustrates the "ongoing" overall firm performance measure index. The second column shows the value of each performance measure in the current quarter (e.g., 2000 Quarter 2). We determine the points for each measure by multiplying the ratio of the measure's current quarter value divided by its baseline value, and we again calculate the weighted points for each measure by multiplying its relative weight by its current points.

Table 7.8 reveals a current quarter index value of 101.9 for the overall firm performance in our illustration. This contrasts with the index value of 102.8 obtained for the distribution index. If this were an actual case rather than a fictitious illustration, one might wonder how to interpret a situation in which the distribution index exceeds the overall firm's index. Basically, this would indicate that the firm is making "relatively" more rapid progress in the subset of performance areas most directly influenced distribution than in its overall performance. Conversely, if the overall performance index exceeds that of the distribution subset, then slower than average progress is being made in the subset area. Clearly, there may be some situations and firms for which these sorts of comparisons have more applicability and/or validity than for others. Nevertheless, having indices for both the overall firm and selected functional areas at least affords the opportunity for this type of comparison.

Table 7.8 Firm Performance Measure Index

Measure		Current Quarter Value [a]	Points	Weighted Points
P1.	ROE (%)	10.5	105.0[b]	7.4[c]
P2.	Market Share (%)	31.0	103.3	7.2
P3.	Cash Flow ($000)	101.0	101.0	6.1
P4.	Sales Growth (%)	11.0	110.0	6.6
P5.	Operating Income ($000)	70.0	106.1	7.4
P6.	Order Fill Rate (%)	85.0	101.2	6.1
P7.	Line Fill Rate (%)	97.0	101.0	7.1
P8.	Number of Customer Partnerships (#)	30	100.0	3.0
P9.	Customer Satisfaction Survey (Rating)	90.0	100.0	7.0
P10.	Manufacturing Cycle Time (Days)	10.0	100.0	7.0
P11.	Cost per Normalized Unit ($/Unit)	0.98	102.0	7.1
P12.	New Product Actual vs. Planned Introduction (Days Past Plan)	39.0	102.5	3.1
P13.	Employee Satisfaction Survey (Rating)	76.0	95.0	3.8
P14.	Order Fulfillment Cycle Time (Days)	10.0	100.0	7.0
P15.	Product Development to Market Cycle Time (Months)	30.0	100.0	7.0
P16.	Percent of Sales from New Products (%)	21.0	105.0	3.1
P17.	Manufacturing Yield (%)	94.0	98.9	5.9
PERFORMANCE INDEX				101.9

[a] **Value in initial period (quarter) of index**

[b] **Sample Calculation: 105.0 = (10.5/10.0) x 100; *i.e.*, Points = current quarter value/baseline value (Table 7.7) x 100**

[c] **Sample Calculation: 7.4 = .07 x105.0; *i.e.*, Weighted Points = Overall Weight (Table 7.3) x Points**

Appendix 7.4B Additional Issues To Consider In Constructing An Index

Two issues need to be considered when constructing a distribution index or any other index using the method described here. The first involves calculating the points for performance measures, such as the average order fulfillment cycle time (P14) and the cost per normalized unit (P11). Generally, a firm wishes to improve its performance in order cycle time by reducing the average time required to fill orders. Similarly, a firm typically seeks to reduce its costs per unit. For performance measures such as these, because the objective is to reduce the measure's value from its initial base value, one must modify the formula to calculate the measure's points. To illustrate this, we consider the "points" calculation for P11, cost per normalized unit. P11 has a baseline index value of 1.00 (refer to Table 7.5) and a current quarter value of 0.98 (refer to Table 7.6). This indicates that the firm has improved (reduced) its operating costs since the base period of the index. Therefore, to account for the performance improvement in P11, we use the following alternative equation to calibrate the points in the current quarter index for P11 (and for all performance measures for which a decrease in the measure's value represents an improvement):

100 - [(current quarter value / baseline value) x 100 - 100], or
200 - [(current quarter value / baseline value) x 100]

One can observe in Table 7.6 that this formula yields a value of 102.0 points for P11, a value which seems reasonably intuitive. Namely, a decrease of 2% in the normalized cost per unit creates a current quarter points value of 102.0. One obvious limitation of this formula is that, by construction, it constrains the points value of a performance measure to a maximum of 200. A formula which calculates the current quarter points of a performance measure, for which an increase in value represents improvement does not have this limit. Obviously one could devise alternative formulas or adjustment mechanisms to eliminate the 200 point ceiling on performance measures where a decrease in value represents improvement. However, for our purposes, the simple formula described above represents a satisfactory approach.

The second issue to consider involves the calculation of "points" in the current quarter for all performance measures. For illustrative purposes, we examine P4 (sales growth) and P7 (line fill rate). These have baseline values of 10% and 96%, respectively (refer to Table 7.5). If sales growth were to increase 20% in some future quarter, then this performance measure would have a current quarter points value of 200. It is certainly plausible that a firm's rate of sales growth could increase from 10% to 20% (or even more) between a baseline period and some future period. Yet, by definition, a firm's line fill rate can at best increase in this example to 100.0 in some future quarter. A fill rate of 100.0 would create a current quarter points value of 104.2 for P7.

As this example illustrates, there are essentially implicit constraints or limitations on the maximum potential points value which a performance measure may reach vis-à-vis another measure and on how quickly one measure may change relative to another. These implicit constraints result from the broad variety of individual performance measures typically found in a balanced scorecard. In constructing an index, therefore, we recommend that one evaluate the differences inherent in a firm's particular set of performance measures. Should this issue represent a concern, one can consider undertaking further steps in the index construction process to address these "implicit constraints". Theoretically, one could develop an adjustment mechanism that essentially "equates" the rate of change in one performance measure to that of another. For example, a scale might equate an increase in line fill rate of .005 (e.g., .97 to .975) to an increase of 5% in sales growth. One could ultimately construct a table or rating scale in which ranges of rates of change for all performance measures could be equated. One would then modify our current set of formulas to incorporate this table. For purposes of simplicity, we do not pursue this discussion further, but, we refer the interested reader back to section 4.3, which illustrated how to employ AHP rating scales. A review of the concept of an AHP rating scale represents a good starting point for the reader interested in incorporating a rate of change "equalization" mechanism into a performance measure index.

7.5 Summary

Measuring supply chain performance and overall performance represents a critical requirement for any firm operating a large-scale network. The common adage that "you can't manage what you can't measure" is certainly a truism. In this chapter, we have reviewed several illustrative hierarchical methods for assessing both supply chain performance and overall firm performance on an ongoing basis.

Section 7.1 offered a hierarchical framework for organizing a complete set of performance measures to monitor all aspects of a firm's operations. This framework facilitates the alignment of performance measures that monitor relatively narrow or "contained" parts of a firm's total operations (e.g., the receiving function at a warehouse) with higher, much broader views and measures of an operation (e.g., a firm's total manufacturing and distribution costs). Further, we observed that a performance measurement system can extend to encompass a total supply chain consisting of multiple firms (e.g., a manufacturer, its suppliers and its customers). Section 7.2 reviewed a number of categories commonly used to group performance measures. Additionally, this section documented many frequently used performance measures. In section 7.3, we reviewed what has become a prominent measure of many firms' total performance, the balanced scorecard. Then in section 7.4, we demonstrated an AHP-based methodology to link quantitatively the performance measures of a balanced scorecard to a firm's mission, objectives and strategies. In particular, we showed how a firm can employ this method to weight the relative importance of its performance measures in terms of overall firm mission and/or specific strategies (e.g., a distribution strategy). With this quantitative link, we were able to develop a composite index of the firm's performance measures. The index facilitates the measurement of the firm's progress in pursuing its overall mission and in tracking the effectiveness of a particular strategy (e.g., the distribution strategy).

Chapter 8

Conclusion

In this book, we have reviewed applications of, and techniques for hierarchical planning across many different components of logistics and supply chain planning. Beginning with a discussion of strategic, tactical and operational planning in the context of a generic hierarchical production planning system in Chapter 1, we then moved on in Chapter 2 to consider a classical example of an HPP system designed by Hax and Meal (1975) for a US tire manufacturer. Our review of HPP implementations in this chapter also included a system installed by a chemical fertilizer manufacturer and a ceramic tile company. Although each HPP implementation was unique, they all shared a number of common characteristics. In particular, each system provided the respective manufacturer with the capability to develop network-wide integrated production and distribution plans spanning the operational, tactical and strategic planning horizons. Further, in each case, the HPP implementations spawned significant improvements in the coordination of the firms' production and distribution activities, and each implementation led to significant reductions in annual operating costs. Each HPP implementation also stimulated an ongoing process of improvement at the respective firms. Specifically, we observed that the benefits of the initial system installations highlighted the potential to reap additional gains at each firm through both enhancements to existing components of the original planning systems and through the development of new components to enhance the original systems.

Chapter 3 focused on hierarchical production planning activities at the individual plant level. In this chapter, we explored how hierarchical production planning techniques are employed to plan and schedule plants in a wide variety of industries and manufacturing processes. This includes process industries, flexible manufacturing systems environments, job shops, fabrication industries and so on. Again, we found common traits in the "plant level" hierarchical planning approaches used in different industries, just as we observed common characteristics in network-wide hierarchical planning techniques used across different industries. Chapter 3 also presented several plant scheduling methodologies that had hierarchical dimensions. Section 3.4 illustrated a model for simultaneously scheduling two linked plants on a manufacturing network; namely, a work-in-process material plant and a finishing plant. This model also incorporated transportation scheduling between the two plants into the planning process. Chapter 3 closed with a brief review of MRP, a classical decision support technique to assist manufacturing planning and scheduling. We observed that MRP has many hierarchical planning dimensions to it, even though one might not immediately think of MRP as an HPP technique.

In Chapter 4, we expanded our discussion from the domain of production to other areas of the supply chain such as distribution and transportation planning.

A review of warehouse location and design planning problems[1], multi-echelon multi-mode transportation problems and a strategic rail network design problem revealed the use and need for hierarchical supply chain planning methods in these areas. All of these problems shared the common trait that they contained a mix of high level questions and decisions (e.g., where to locate a warehouse) and more detailed questions (e.g., how to design the pick area of a warehouse). This mix demanded a multi-level decision-making methodology such as hierarchical planning. We next considered the subject of short run operational transportation scheduling. This discussion illustrated "hierarchical issues" such as that detailed data regarding actual daily demands, transportation loads and deliveries is required to facilitate transportation scheduling. This contrasts with the higher level "average" demand and delivery quantity data that forms the basis for tactical and strategic transportation planning. We observed that to design and maintain an effective transportation network, it is necessary to establish feedback links that convey operational transportation scheduling considerations back to the tactical and in some cases even the strategic levels. Through this feedback, potential infeasibilities or operational problems that are masked at the tactical or strategic levels by data stated in "averages" become known and can be addressed.

Chapter 4 also highlighted another "hierarchical dimension" encountered in transportation scheduling, as well as other areas. Mathematical models to solve transportation scheduling problems (e.g., the traveling salesman problem) are often extremely complex, and in some cases cannot be solved directly (e.g., "np hard" mathematical problems). This can necessitate using algorithmic techniques that utilize various decomposition approaches to solve the scheduling problem. We observed that a decomposition approach involves breaking one mathematical problem into a series of (hierarchical) sub-problems that are mathematically tractable. Results from these sub-problems are used collectively to generate an estimated or "near optimal" solution to the original single problem. Decomposition techniques permeate scheduling and planning problems in virtually all areas of supply chain operations (including transportation), and represent yet another "hierarchical" characteristic of supply chain planning.

Chapter 4 also introduced another hierarchical modeling methodology that one can employ to support logistics planning; namely, the analytic hierarchy process (Saaty, 1996). The application presented in section 4.3 demonstrated how planners can use the AHP to assist in making long-run, tactical transport mode choice decisions on a network. In particular, we illustrated how the AHP could help quantify and weight "qualitative" decision factors into one unified decision support framework that collectively weighed all decision factors, both quantitative and qualitative, in a transport mode and carrier selection problem. In practice, we observed that one can employ the AHP methodology in a broad array of supply chain decisions (and we did in fact utilize the AHP again in Chapter 7 to develop a performance measurement system). Finally, in Chapter 4, we also examined a methodology available in the commercial marketplace to support effective transportation negotiating by shippers. This system known as OptiBid incorporates

[1] Warehouse planning was reviewed in more detail in Chapter 5.

mathematical optimization into a hierarchical process that examines both the individual shipper's transport flows, as well as the flows and infrastructure of multiple carriers' networks to assist in determining the best transport partners for an individual shipper.

In Chapter 5, we reviewed how the design and management of warehouses benefits from a hierarchical planning approach. Decisions about a network of distribution centers or warehouses range from such high level questions as to the optimal number, locations and technologies to use on a warehouse network to such operational questions as basic warehouse operating policies (e.g., what are the best putaway/storage operating procedures). In between these types of strategic and operational decisions reside such tactical activities as annual warehouse capacity planning, the continual balancing of demands against warehouse capacity across the network, and so on. The mix of warehousing decisions reviewed made it clear that an effective warehouse management approach must integrate the impact of potential decisions at each level on the other levels of the planning and operational hierarchy. Chapter 5 also detailed the importance of integrating the internal operations[2] of the warehouses on a network with other key supply chain activities such as manufacturing and distribution. Section 5.2 demonstrated techniques for integrating decision support tools such as DC workload models with the forecasting, inventory management and manufacturing planning and scheduling systems that drive a firm's supply chain. In this section, we further reviewed the versatility of such tools as the DC workload model which can play a role in both operational and tactical planning and scheduling activities. Finally, Chapter 5 presented a noteworthy hierarchical warehouse modeling methodology that Gray, Karmarkar and Seidmann (1992) implemented to guide the warehouse operations of an automotive manufacturer and distributor.

Chapter 6 addressed inventory management and forecasting, two key interwoven elements of supply chain management. A review of the broad spectrum of inventory management decisions that must be made in operating a manufacturing and distribution network demonstrated the need for a hierarchical planning approach in this area. We observed that inventory decisions range from determining the optimal overall level of total investment at the strategic level, to product family investment and deployment decisions at the tactical level, to sku investment and deployment decisions at the operational level. Further, we observed that inventory investment decisions made at each planning level significantly affected decisions at the next lower planning level. This again highlighted the need for feedback between planning levels. Sections 6.2 and 6.3 considered hierarchical methods used in practice to manage inventory on networks. This included ABC classification schemes for grouping sku's, setting desired service level fill rates and inventory targets, and so on. Additionally, we presented an algorithm that aggregated inventory levels of individual end items into product family inventory levels for annual/tactical production and inventory planning purposes. This algorithm illustrated that in order to develop accurate product family data for tactical planning (in this case to develop beginning of planning

[2] *i.e.*, operations within the four walls of the warehouses on a network.

horizon inventory levels), it is sometimes necessary to evaluate detailed item data that is more commonly used to support operational planning and scheduling activities.

The second half of Chapter 6 focused on developing forecasts to support supply chain operations. We began by reviewing the numerous hierarchical dimensions that are frequently used to develop forecasts such as product structure, geographic hierarchies, sales organization hierarchies and so on. Next, we discussed top-down and bottom-up forecasting methods and the potential benefits of a hierarchical forecasting approach. This led into a discussion of the Sales and Operations Planning process that has become widely used in private industry to assist in coordinating supply chain planning activities with overall business requirements. Our discussion of the S&OP process highlighted that this represents a very hierarchical monthly process that includes a formal schedule of reviews and meetings between various functional areas of the firm. The process integrates the plans of each major function in a business (e.g., sales, marketing, manufacturing, demand management, etc...) to form one unified business and supply chain plan. Each functional area shares its latest plans which are eventually amalgamated at the monthly S&OP meeting. Decisions made in this final meeting are then fed back to each functional and/or geographic area so that each area can move forward with the assurance that its planned actions will mesh with those of the other areas of the company. We observed that the S&OP process takes on a dramatically increased level of complexity when a firm operates this process at a global level (as opposed to operating the S&OP at a national level). The hierarchical requirements of aggregating country level demands and businesses to a global perspective necessitates new levels of aggregation, disaggregation and feedback.

Chapter 7 offered several frameworks for maintaining measures to monitor supply chain and total business performance. We observed that a hierarchical framework for performance measurement facilitates the development of a supply chain-wide measurement system in which both small and large functional areas can have meaningful participation. Additionally, the hierarchical approach allows a firm to easily review its complete set of measurements to ascertain whether it has consistent goals and objectives across its total organization.

In conclusion, there exist numerous hierarchical attributes of supply chain decision-making that are common across individual functional areas of the supply chain. The roots of hierarchical planning for large scale networks may have started in the form of hierarchical production planning[3]; however, it is clear that this approach has no functional or process boundaries. By its very nature, virtually any planning problem and/or system that spans a horizon of the short run (several days, weeks or quarters) to the long run (multiple years) must adopt a hierarchical approach. This applies to all aspects of logistics and supply chain planning, as well as realms far beyond the supply chain. In this book, we have illustrated this concept by examining planning and scheduling requirements and techniques in a diverse array of logistics activities. Given that firm activities on a supply chain really should represent one integrated, continuous process; it is not surprising to

---

[3] See discussion of Hax and Meal (1975) in Chapter 2.

find common methodologies (*i.e.*, hierarchical planning approaches) applicable and prevalent across individual functional areas of the process.

In closing this book, we return to our opening discussion in Chapter 1 on the rationale for employing a hierarchical planning approach. In each of the areas explored in Chapters 1 through 7, it is clear that one cannot build single monolithic planning models to provide decision support for strategic, tactical and operational planning. The multitude of decisions spanning multiple planning horizons are far to complex for one single planning model to address. Highly detailed plans and schedules are required at the operational planning level, while at the strategic level, highly summarized data views and plans are needed. The mathematical complexity of decision support models required at each planning level also precludes the possibility of developing single models to support all planning activities. Even if one can develop such models conceptually; one generally cannot solve them mathematically. Instead a decomposition approach utilizing multiple sub-models is required.

We also observed the common theme that models at each level should be tailored to the planning needs of the managers and employees operating at a particular level of the organization. Senior executives charged with responsibility for long run strategy have significantly different decision support requirements than operational colleagues responsible for the short run scheduling of production lines, distribution centers and so on.

The risk and timing of decisions also dictates a hierarchical approach. In all functional areas of the supply chain, the timing and risk of strategic decisions differs greatly from that of tactical and operational decisions. The difference in the risk and timing considerations of a strategic decision to build a plant versus the operational decision to schedule a production line for two weeks is self-evident. The plant decision bears enormous risk and typically requires multiple years to implement, while the line scheduling decision is relatively low risk and the results of such a decision become clear rather quickly.

The critical importance of well-designed feedback loops from the operational level to the tactical level to the strategic level represents another overriding theme observed across supply chain operations. Whether the functional area is forecasting, inventory management or warehousing; a planning system that only has top-down information flows will not facilitate effective, efficient operations. We observed that decisions that appear feasible at a high level (e.g., production line schedules generated by a network-wide optimization model) may be infeasible at a lower level. The feedback loops that are the hallmark of hierarchical systems allow more successful decision-making at all levels by conveying infeasibilities and operating issues back from lower planning and operating levels to higher planning levels.

In conclusion, hierarchical planning systems require significant time and resources to develop and maintain. However, the return on investing in this planning and scheduling methodology more than justifies the effort and cost.

References

Aardal, K.and Larsson, T. (1990), A Benders Decomposition Based Heuristic For The Hierarchical Production Planning Problem, *European Journal Of Operations Research*, 45, 4-14.

Ackoff, R.L. (1970), A *Concept Of Corporate Planning*, New York, Wiley Interscience.

van Amstel, R.P. and D'hert, G. (1996), Performance Indicators In Distribution, *The International Journal Of Logistics Management*, 7(1), 73-82.

Anderson, H., Axsater, S., and Jonsson, H. (1980), A Simulation Study Of Hierarchical Production-Inventory Control, *OR Spektrum*, 2(2), 79-89.

Anthony, R.N. (1965), *Planning And Control Systems: A Framework For Analysis*, Boston, Harvard University.

APICS Dictionary (1999), American Production And Inventory Control Society, Alexandria, Va.

Axsater, S. (1981), Aggregation Of Product Data For Hierarchical Production Planning, *Operations Research*, 30, 744-755.

Axsater, S. (1979), On The Design Of The Aggregate Model In A Hierarchical Production Planning System, *Engineering And Process Economics*, 4, 89-97.

Axsater.S., and Nuttle, H.L.W. (1986), Aggregating Items In Multi-Level Lot Sizing, in *Multi-Stage Production Planning And Inventory Control,* Lecture Notes In Economics And Mathematical Systems, Axsater, S., Schneeweiss, C. and Silver, E. (eds)., Heidelberg, Springer-Verlag.

Axsater, S. and Nuttle, H.L.W. (1987), Combining Items For Lot Sizing In Multi-Level Assembly Systems, *International Journal Of Production Research*, 25(6), 795-807.

Axsater, S., and Rosling, K. (1994), Multi-Level Production-Inventory Control: Material Requirements Planning Or Reorder Point Policies?, *European Journal Of Operational Research*, 75(2), 405-412.

Belarmino, A-D., Gonzalez, M., and Garcia, E. (1998), A Hierarchical Approach To Managing Dairy Routing, *Interfaces*, 28(2), 21-31.

Bitran, G.R., Haas, E.A., and Hax, A.C. (1981), Hierarchical Production Planning: A Single Stage System, *Operations Research*, 30, 717-743.

Bitran, G.R., Haas, E.A., and Hax, A.C. (1982), Hierarchical Production Planning: A Two-Stage System, *Operations Research*, 31, 232-251.

Bitran, G.R. and Hax, A.C. (1977), On The Design Of Hierarchical Production Planning Systems, *Decision Sciences*, 8, 28-55.

Bitran, G.R., and Tirupati, D. (1993), Hierarchical Production Planning, *in Logistics Of Production And Inventory*, Graves, S.C., Rinnooy Kan, A.H.G., and Zipkin, P.H., *Series Handbooks In Operations Research And Management Sciences*, Vol. 4, Amsterdam, North-Holland, 523-568.

Bonney, M.C., Tien, M.A., Huang, C.C., and Barson, R.J. (1996), Inventory And Enterprise Integration, *International Journal Of Production Economics*, 45 (1-3), 91-99.

Bowers, M.R., and Jarvis, J.P. (1992), A Hierarchical Production Planning And Scheduling Model, *Decision Sciences*, 23(1), 144-159.

Bowersox, D.J., and Closs, D.J. (1996), *Logistical Management, The Integrated Supply Chain Process*, New York, McGraw-Hill.

Bowersox, D.J., Daugherty, P., Droge, C., Rogers, D. and Wradlow, D. (1989), *Leading Edge Logistics Competitive Positioning For The 1990's*, Oakbrook, IL., Council Of Logistics Management.

Brown, G.G., Ellis, C.J., Graves, G.W. and Ronen, D. (1987), Real-Time, Wide Area Dispatch Of Mobil Tank Trucks, *Interfaces*, 17(Jan-Feb), 107-120.

Burch, E.E., Oliff, M.D., and Sumichrast, R.T. (1987), Linking Level Requirements In Production Planning And Scheduling, *Production And Inventory Management Journal*, 2nd quarter, 123-131.

Carravilla, M.A., and de Sousa, J.P. (1995), Hierarchical Production Planning In A Make-To-Order Company, *European Journal Of Operational Research*, 86(1), 43-56.

Carter, M.W., Farvolden, J.M., Laporte, G. and Xu, J. (1996), Solving An Integrated Logistics Problem Arising In Grocery Distribution, *INFOR*, 34(4), 290-306.

Cheung, R.K. and Powell, W.B. (1996), An Algorithm For Multistage Dynamic Networks With Random Arc Capacities, With An Application To Dynamic Fleet Management, *Operations Research*, 44(6), 951-963.

Chuan, J.C., Lasserre, J.B., and Roubellat, F. (1981), Hierarchical Planning: A Case Study, in *Operational Research* '81, J.B. Brans, (ed), Amsterdam, North-Holland Publishing Company.

Chung, C-H., and Krajewski, L.J. (1987), Interfacing Aggregate Plans And Master Production Schedules Via A Rolling Horizon Feedback Procedure, *Omega*, 15(5), 401-409.

Clark, A.R. and Armentano, V.A. (1995), A Heuristic For A Resource-Capacitated Multi-Stage Lot Sizing Problem With Lead Times, *Journal Of The Operational Research Society*, 46(10), 1208-1222.

Clark, G. and Wright, J. (1964), Scheduling Of Vehicles From A Central Depot To A Number Of Delivery Points, *Operations Research*, 12(4), 568 – 581.

Copacino, W.C. (1999), A New Direction In Strategic Planning, *Logistics Management And Distribution Report*, 38 (9), 38.

Copacino, W.C. (1996), Why You Should Synchronize Logistics And Manufacturing, *Logistics Management*, 35 (11), 63.

Coyle, J.J. (1995), Research For Hire, *Transportation And Distribution*, 36(2), 64 - 68.

Crainic, T.G. and Roy, J. (1988), OR Tools For Tactical Freight Transportation Planning, *European Journal Of Operational Research*, 33(3), 290-297.

Daganzo, C.F. (1987), The Breakbulk Role Of Terminals In Many-To-Many Logistics Networks, *Operations Research*, 35(4), 543-555.

Dangerfield, B.J. and Morris, J.S. (1992), Top-Down Or Bottom-Up: Aggregate Versus Disaggregate Extrapolations, *International Journal Of Forecasting*, 8, 233 - 241.

Dantzig, G.B., Fulkerson, R. and Johnson, S. (1954), Solution Of A Large Traveling Salesman Problem, *Operations Research*, 2(4), 393 – 410.

Dantzig, G.B. and Ramser, J.H. (1959), The Truck Dispatching Problem, *Management Science*, 6(1), 80 – 91.

Dean, B.V. and Nishry, M. (1965), Scoring And Profitability Models For Evaluation And Selecting Engineering Projects, *Operations Research*, 13, 550-569.

Dempster, M.A., Fisher, M.L., Jansen, B.J., Lageweg, J.K., Lenstra, J.K. and Rinnooy Kan, A.H.G. (1983), Analysis Of Heuristics For Stochastic Programming:

Results For Hierarchical Scheduling Problems, *Mathematics Of Operations Research*, 8(4), 525-537.

De Matta, R. and Miller, T.C. (1993), A Note On The Growth Of A Production Planning System: A Case Study In Evolution, *Interfaces*, 23(4), 116-122.

De Matta, R. and Miller, T.C. (2000), An Integrated Short-Term Production And Distribution Scheduling Problem On A Two Echelon Manufacturing Network, *Working Paper*.

De Matta, R., and Miller, T.C. (1996), Evaluating The Trade-Offs Between Idling And Shutting Down Production Lines In Process Industries, *Production And Operations Management*, 5(4), 371-390.

Desrochers, M., Lenstra, J.K. and Savelsbergh, M.W.P. (1990), A Classification Scheme For Vehicle Routing And Scheduling Problems, *European Journal Of Operational Research*, 46, 322-332.

Dillman, R., Becker, B. and Beckefeld, V. (1996), Practical Aspects Of Route Planning For Magazine And Newspaper Wholesalers, *European Journal Of Operational Research*, 90, 1-12.

Earl, B.E., and Sumichrast, R.T. (1987), Using Hierarchical Techniques To Schedule Machines With Simultaneous Production Capabilities, in *Applications Of Management Science*, vol. 5, Schultz, R.L. (ed.), Amsterdam, North-Holland Publishing Company.

Erschler, J., Fontan, G., and Merce, C. (1986), Consistency Of The Disaggregation Process In Hierarchical Planning, *Operations Research*, 34(3), 464-469.

Evans, J.R., Anderson, D.R., Sweeney, D.J., and Williams, T.A. (1987), *Applied Production And Operations Management*, New York, West Publishing Company.

Fawcett, S.E. and Cooper, M.B. (1998), Logistics Performance Measurement And Customer Success, *Industrial Marketing Management*, 27, 341-357.

Fisher, M.L. and Jaikumar, R. (1981), A Generalized Assignment Heuristic For Vehicle Routing, *Networks*, 11(2), 109 – 124.

Fontan, G., Imbert, S. and Merce, C. (1985), Consistency Analysis In A Hierarchical Production Planning System, *Engineering Costs And Production Economics*, 9, 193-199.

Foreman, E.H., Saaty, T.L., Selley, M.A., and Whitaker, R. (1990), *Expert Choice*, Pittsburgh, PA., Expert Choice Inc.

Franca, P.M., Armentano, V.A., Berretta, R.E., and Clark, A.R. (1997), A Heuristic Method For Lot-Sizing In Multi-Stage Systems, *Computers and Operations Research*, 24(9), 861-874.

Fransoo, J.C., Sridharan, V., and Bertrand, J.W.M (1995), A Hierarchical Approach For Capacity Coordination In Multiple Products Single-Machine Production Systems With Stationary Stochastic Demands, *European Journal Of Operational Research*, 86(1), 57-72.

Gabbay, H. (1979), Optimal Aggregation And Disaggregation In Hierarchical Planning, in *Disaggregation: Problems In Manufacturing And Service Organizations*, Rizman, L.R., Krajewski, L.J., Berry, W.L., Goodman, S.H., Hardy, S.T. and Vitt, L.D. (eds), Boston, Martinus Nijhoff Publishing.

Gelders, L.F., and van Steelandt, F.V. (1980), Design And Implementation Of A Production Planning System In A Rolling Mill: A Case Study, *AIIE Transactions*, 12(1), 54-58.

Gelders, L.F., and van Wassenhove, L.N. (1982), Hierarchical Integration In Production Planning: Theory And Practice, *Journal Of Operations Management*, 3(1), 27-35.

Ghosh, S. and Gaimon, C. (1992), Routing Flexibility And Production Scheduling In A Flexible Manufacturing System, *European Journal Of Operational Research*, 60(3), 344-364.

Glover, F., Jones, G., Karney, D., Klingman, D. and Mote, J. (1979), An Integrated Production, Distribution, And Inventory Planning System, *Interfaces*, 9(5),

Golden, B.L. and Assad, A.A. (1988), *Vehicle Routing: Methods And Studies*, New York, North-Holland.

Golden, B.L., Assad, A.A. and Dahl, R.W. (1984), Analysis Of A Large Scale Vehicle Routing Problem With An Inventory Component, *Large Scale Systems*, 7(2 and 3), 181-190.

Golden, B.L., Laporte, G. and Taillard, E.D. (1997), An Adaptive Memory Heuristic For A Class Of Vehicle Routing Problems With Minimax Objectives, *Computers And Operations Research*, 24(5), 445 – 452.

Golden, B.L. and Yee, J., (1979) A Framework For Probabilistic Vehicle Routing, *AIIE Transactions*, 11, 109 - 112.

Golden, B.L., Wasil, E.A., and Levy, D.E., Application Of The Analytic Hierarchy Process: A Categorized, Annotated Bibliograpy, in *The Analytic Hierarchy*

Process: Applications And Studies, Golden, B.L., Wasil, E.A., and Harker, P.T., (eds.), Berlin, Springer-Verlag.

Golenko-Ginzburg, D., Sinuany-Stern, Z., and Kats, V. (1996), A Multilevel Decision-Making System With Multiple Resources For Controlling Cotton Harvesting, *International Journal Of Production Economics*, 46-47 (Dec), 55-63.

Gordon, T.P., Morris, J.S., and Dangerfield, B.J. (1997), Top-Down Or Bottom-Up: Which Is The Best Approach To Forecasting ?, *The Journal Of Business Forecasting*, (Fall), 13 - 16.

Graves, S.C. (1981), Multi-Stage Lot-Sizing: An Iterative Procedure, Chapter 4 in *Multi-Level Production/Inventory Control Systems: Theory And Practice,* Schwarz, L.B. (ed.), Amsterdam, North-Holland Publishing Company.

Gray, A.E., Karmarkar, U.S., and Seidmann, A. (1992), Design And Operation Of An Order-Consolidation Warehouse: Models And Application, *European Journal Of Operational Research*, 58, 14 - 36.

Harker, P.T., and Vargas, L.G. (1990), Reply To 'Remarks On The Analytic Hierarchy Process by J.C. Dyer, *Management Science*, 36 (3), 269 - 273.

Hax, A.C., and Bitran, G.R. (1979), Hierarchical Production Planning Systems: A Production Application, in *Disaggregation Problems In Manufacturing And Service Organizations*, Rizman, L.R., Krajewski, L.J., Berry, W.L., Goodman, S.H., Hardy, S.T., and Vitt, L.D., (eds), Boston, Martinus Nijhoff Publishing.

Hax, A.C., and Golovin, J.J. (1978), Hierarchical Production Planning Systems, in *Studies In Operations Management*, Hax, A.C., (ed), Amsterdam, North-Holland.

Hax, A.C. and Meal, H.C. (1975), Hierarchical Integration Of Production Planning And Scheduling, in M.A. Geisler (ed.), *TIMS Studies In Management Science, Vol. 1, Logistics*, Amsterdam, North-Holland.

Hax, A.C. (1977), Integration Of Strategic And Tactical Planning In The Aluminum Industry, in *Applied Mathematical Programming*, Bradley, S.P., Hax, A.C., and Magnanti, T.L., (eds), Reading, Addison-Wesley.

Hill, J.M. (1996), Warehouse Management Systems: The Competitive Tool For The 1990's: Part One, *Modern Materials Management*, 51(11), W3-W16.

Ho, P-K. and Perl, J. (1995), Warehouse Location Under Service-Sensitive Demand, *Journal Of Business Logistics*, 16(1), 133-162.

House, R.G. and Jamie, K.D. (1981), Measuring The Impact Of Alternative Market Classification Systems In Distribution Planning, *Journal of Business Logistics*, 2, 1-31.

Jaikumar, R. and Van Wassenhove, L.N. (1989), Production Planning In Flexible Manufacturing Systems, *Journal Of Manufacturing And Operations Management*, 2, 52-78.

Johnson, M.E., and Davis, T. (1998), Improving Supply Chain Performance By Using Order Fulfillment Metrics, *National Productivity Review/Summer 1998*, New York, John Wiley & Sons Inc.

Kahn, K.B. (1998), Revisiting Top-Down Versus Bottom-Up Forecasting, *Journal Of Business Forecasting*, Summer, 14-19.

Kaplan, R.S. and Norton, D.P. (1992), The Balanced Scorecard - Measures That Drive Performance, *Harvard Business Review*, Jan.- Feb., 71-79.

Kaplan, R.S. and Norton, D.P. (1993), Putting The Balanced Scorecard To Work, *Harvard Business Review*, Sept.- Oct., 134-147.

Kaplan, R.S. and Norton, D.P. (1996), Using The Balanced Scorecard As A Strategic Management System, *Harvard Business Review*, Jan.- Feb., 75-85.

Katayama, H. (1996), On A Two-Stage Hierarchical Production Planning System For Process Industries, *International Journal Of Production Economics*, 44(1,2), 63-72.

Keefer, D.L. (1978), Allocation Planning For R&D With Uncertainty And Multiple Objectives, *IEEE Transactions On Engineering Management*, 25(1), 8 -14.

Keown, A.J., Taylor, B.W. and Duncan, C.P. (1979), Allocation Of Research And Development Funds: A Zero-One Goal Programming Approach, *Omega*, 7, 345-351.

Kok, A.G., de. (1990), Hierarchical Production Planning For Consumer Goods, *European Journal Of Operations Research*, 45, 55-69.

Koskosidis, Y.A. and Powell, W.B. (1990), Application Of Optimization Based Models On Vehicle Routing And Scheduling Problems With Time Window Constraints, *Journal Of Business Logistics*, 11(2), 101-128.

Krajewski, L.J., and Rizman, L.P. (1977), Disaggregation In Manufacturing And Service Organizations: Survey Of Problems And Research, *Decision Sciences*, 8, 1-18.

Lambert, D.M. and Sharma, A. (1990), A Customer-Based Competitive Analysis For Logistics Decisions, *International Journal Of Physical Distribution And Materials Management*, 20 (1), 17 - 24.

Lambert, D.M. and Cook, R.L. (1990), *Integrating Marketing And Logistics For Profit*, Business, 40 (3), 22 - 29.

Langley, J.C., and Holcomb, M.C. (1992), Creating Logistics Customer Value, *Journal Of Business Logistics*, 13 (2), 1 - 27.

Langley, J.C. (1992), The Evolution Of The Logistics Concept, in *Logistics: The Strategic Issues*, Martin, C. (ed.), London, Chapman And Hall, 17 - 28.

Lapide, L. (1998), A Simple View Of Top-Down Versus Bottom-Up Forecasting, *The Journal Of Business Forecasting*, Summer, 28-29.

Laporte, G. (1992), The Vehicle Routing Problem: An Overview Of Exact And Approximate Algorithms, *European Journal Of Operational Research*, 59, 345 – 358.

Lasserre, J.B., Martin, J.P. and Roubellat, F. (1983), Aggregate Model And Decomposition Method For Mid-Term Production Planning, *International Journal Of Production Research,* 21(6), 835 - 843.

Leong, G.K., Oliff, M.D., and Markland, R.E. (1989), Improved Hierarchical Production Planning, *Journal Of Operations Management*, 8(2), 90-114.

Liberatore, M.J., and Miller, T.C. (1995), A Decision Support Approach For Transport Carrier And Mode Selection, *Journal Of Business Logistics*, 16(2), 85 - 116.

Liberatore, M.J., and Miller, T.C. (1998), A Framework For Integrating Activity-Based Costing And The Balanced Scorecard Into The Logistics StrategyDevelopment And Monitoring Process, *Journal Of Business Logistics,* 19(2), 131 - 154.

Liberatore, M.J., and Miller, T.C. (1985), A Hierarchical Production Planning System, *Interfaces*, 15(4), 1-11.

Liberatore, M.J., Monahan, T.F., and Stout, D.E. (1992), A Framework For Integrating Capital Budgeting Analysis With Strategy, *The Engineering Economist*, 38(1), 31-43.

Liberatore, M.J., Monahan, T.F., and Tymon, W.G. (1997), A Framework For Integrating A Balanced Scorecard To Company Strategy, *Corporate Controller*, Winter, 24-29.

Liberatore, M.J., Nydick, R.L., and Sanchez, P. (1992), The Evaluation Of Research Papers (Or How To Get An Academic Committe To Agree On Something), *Interfaces*, 22(2), 92 - 100.

Lin, C-W., and Moodie, C.L. (1989), Hierarchical Production Planning For A Modern Steel Manufacturing System, *International Journal Of Production Research*, 27(4), 613-628.

Ling, R.C., and Goddard, W.E. (1988), *Orchestrating Success*, New York, NY, John Wiley & Sons Inc.

Maxwell, W., Muckstadt, J.A., Thomas, L.J., and VanderEecken, J. (1983), A Modeling Framework For Planning And Control Of Production In Discrete Parts Manufacturing And Assembly Systems, *Interfaces*, 13(6), 92-104.

Meal, H.C. (1978), A Study Of Multi-Stage Production Planning, *in Studies In Operations Management*, Hax, A.C., (ed), 253-285, Amsterdam, North-Holland.

Meal, H.C. (1984), Putting Production Decisions Where They Belong, *Harvard Business Review*, March-April, 102-110.

Meal, H.C., Wachter, M.H., and Whybark, D.C. (1987), Material Requirement Planning In Hierarchical Production Planning Systems, *International Journal Of Production Research,* 25(7), 947-956.

Melnyk, S.A., Gonzalez, R.F., and Anderson, S.J. (1983), Manufacturing Resource Planning: Insights Into A New Corporate Way Of Life, *APICS Research Report*, American Production And Inventory Control Society, Report 82-8.

Mentzer, J.T. and Konrad, B.P. (1991), An Efficiency/Effectiveness Approach To Logistics Performance Analysis, *Journal Of Business Logistics*, 12(1), 33-62.

Meybodi, M.Z. (1995), Integrating Production Activity Control Into A Hierarchical Production Planning Model, *International Journal Of Operations And Production Management*, 15(5), 4-25.

Miller, T.C. and Liberatore, M.J. (1988), Implementing Integrated Production And Distribution Planning Systems, *International Journal Of Operations And Production Management*, 8(7), 31-41.

Miller, T.C. and Liberatore, M.J. (1989), Production Planning In a Process Firm, *Production And Inventory Management Journal*, 1st quarter, 44-48.

Miller, T.C. (1986), A Simplified Hierarchical System Linking An Annual Production Model With A Weekly Master Production Schedule, *Proceedings Of*

The 22nd Annual Meetings Of The Institute Of Management Sciences, Southeastern Chapter, Norfolk, VA, Old Dominion University.

Miller, T.C. (1991), Integrating Current End-Item Inventory Conditions Into Optimization-Based Long-Run Aggregate Production And Distribution Planning, *Production And Inventory Management Journal*, 4th quarter, 74-80.

Miller, T.C. (1991A), The International Modal Decision, *Distribution*, 90(11), 82-92.

Miller, T.C. (1991B), Air vs. Ocean: Two Critical Factors, *Distribution*, 90(12), 46-52.

Miller, T.C. (1993), *Transport Cost Analysis: Finding The Right Mode*, Allegan, Michigan, Boskage Commerce Publishers.

Miller, T.C., Wise, D., and Clair, L. (1996), Transport Network Design And Mode Choice Modeling For Automobile Distribution: A Case Study, *Location Science*, 4(1/2), 37-48.

Mohanty, R.P. and Krishnaswamy, M.V.R. (1984), An Assessment Of Some Approaches To Hierarchical Production Planning, *International Journal Of Operations And Production Management*, 4(1), 48-60.

Mohanty, R.P. and Kulkarni, R.V. (1987), Hierarchical Production Planning: Comparison Of Some Heuristics, *Engineering Costs And Production Economics*, 11, 203-214.

Mohanty, R.P., and Singh, R. (1992), A Hierarchical Production Planning Approach For A Steel Manufacturing System, *International Journal Of Operations And Production Management*, 12(5), 69-78.

Kaplan, R.S., and Norton, D.P. (1992), The Balanced Scorecard - Measures That Drive Performance, *Harvard Business Review*, 70 (Jan-Feb), 71 - 79.

Oeltjenbruns, H., Kolarik, W.J., and Schnadt-Kirschner, R. (1995), Strategic Planning In Manufacturing Systems – AHP Application To An Equipment Replacement Decision, *International Journal Of Production Economics*, 38(2,3), 189-197.

Oliff, M.D. (1987), Disaggregate Planning For Parallel Processors, *IIE Transactions*, 19(2), 215-221.

Oliff, M.D., and Burch, E.E., (1985), Multiproduct Production Scheduling At Owens-Corning Fiberglas, *Interfaces*, 15(5), 25-34.

OptiBid, (1999), *OptiBid - Optimized Bidding And Lane Assignment Software System*, The Sabre Group, Burlington, MA.

Orlicky, J. (1975), *Material Requirements Planning*, New York, NY, McGraw-Hill Book Company.

Ozdamar, L., Bozyel, M.A., and Birbil, S.I., (1998), A Hierarchical Decision Support System For Production Planning, *European Journal Of Operational Research*, 104(3), 403-422.

Partovi, F.Y. and Hopton, W.E., (1994), The Analytic Hierarchy Process As Applied To Two Types Of Inventory Problems, *Production And Inventory Management Journal*, 35(1), 13-19.

Perl, J.S and Daskin, M.S., (1985), A Warehouse Location-Routing Problem, *Transportation Research B*, 19B(5), 381-396

Plossl, G.W., (1980), MRP Yesterday, Today And Tomorrow, *Production And Inventory Management Journal*, 21, 1-10.

Powell, W.B. and Mayoras, D.E., (1997), Finding The Yellow Brick Road: Part 5, Through The Chyrstal Ball, *Interfaces*, 27(3), 14 – 21.

Powell, W.B., Sheffi, Y., Nickerson, K.S., Butterbaugh, K. and Atherton, S., (1988), Maximizing Profits For North American Van Lines Truckload Division: A New Framework For Pricing And Operations, *Interfaces*, 18(1), 21 – 41.

Powell, W.B. and Sheffi, Y., (1986), Interactive Optimization For Motor Carrier Load Planning, *Journal Of Business Logistics*, 7(2), 64-90.

Powell, W.B. and Sheffi, Y., (1989), Design And Implementation Of An Interactive Optimization System For Network Design In The Motor Carrier Industry, *Operations Research*, 37(1), 12-29.

Rogers, D. F., Plante, R.D., Wong, R.T., and Evans, J.R., (1991), Aggregation And Disaggregation Techniques And Methodology In Optimization, *Operations Research*, 39(4), 553-582.

Ronen, D., (1988), Perspectives On Practical Aspects Of Truck Routing And Scheduling, *European Journal Of Operational Research*, 35, 137 – 145.

Rosling, K., (1981), Optimal Lot-Sizing For Dynamic Assembly Systems, Linkoping Institute Of Technology, Sweden.

Saad, G.H., (1990), Hierarchical Production-Planning Systems: Extensions And Modifications, *Journal Of The Operational Research Society*, 41(7), 609-624.

Saaty, T.L. (1996), *The Analytic Hierarchy Process*, Pittsburgh, PA, RWS Publications.

Savelsbergh, M.W.P. and Goetschalckx, M., (1995), A Comparison Of The Efficiency Of Fixed Versus Variable Vehicle Routes, *Journal Of Business Logistics*, 16(1), 163 – 187.

Segerstedt, A. (1996), A Capacity-Constrained Multi-Level Inventory And Production Control Problem, *International Journal Of Production Economics,* 45(1-3), 449-461.

Shapiro, J.F. (1984), A Note On Node Aggregation And Benders Decomposition, *Mathematical Programming*, 29 (1), 113 - 119.

Shapiro, J.F., Singhal, V.M., and Wagner, S.N. (1993), Optimizing The Value Chain, *Interfaces*, 23 (2), 102 - 117.

Silver, E., (1990), Deliberately Slowing Down Output In A Family Production Context, *International Journal Of Production Research*, 28(1), 17-27.

Sodhi, M.S., Askin, R.G., and Sen, S., (1994), A Hierarchical Model For Control Of Flexible Manufacturing Systems, *Journal Of The Operational Research Society*, 45(10), 1185-1196.

Sohner, V., and Schneeweiss, C., (1995), Hierarchically Integrated Lot Size Optimization, *European Journal Of Operational Research*, 86(1), 73-90.

Solomon, M.M., (1986), On The Worst-Case Performance Of Some Heuristics For The Vehicle Routing And Scheduling Problem With Time Window Constraints, *Networks*, 16, 161 – 174.

Solomon, M.M., (1987), Algorithms For The Vehicle Routing And Scheduling Problem With Time Window Constraints, *Operations Research*, 35, 254 – 269.

Srivastava, B., and Chen, W-H. (1996), Batching In Production Planning For Flexible Manufacturing Systems, *International Journal Of Production Economics*, 43(2,3), 127-137.

Stadtler, H., (1986), Hierarchical Production Planning: Tuning Aggregate Planning With Sequencing And Scheduling, in *Multi-Stage Production Planning And Inventory Control, Lecture Notes In Economics And Mathematical Systems*, Axsater, S., Schneeweiss, C., and Berlin, E.S., Heidelberg, Springer-Verlag.

Stadtler, H., (1996), Mixed Integer Programming Model Formulations For Dynamic Multi-Item Multi-Level Capacitated Lotsizing, *European Journal Of Operational Research*, 94(3), 561-581.

Takahashi, K., Hiraki, S., and Soshiroda, M., (1994), Pull-Push Integration In Production Ordering Systems, *International Journal Of Production Economics*, 33(1-3), 155-161.

Tang, C.S., and Denardo, E.V., (1988A), Models Arising From A Flexible Manufacturing Machine, Part 1: Minimization Of The Number Of Tool Switches, *Operations Research*, 36, 767-777.

Tang, C.S., and Denardo, E.V., (1988B), Models Arising From A Flexible Manufacturing Machine, Part 2: Minimization Of The Number Of Switching Instants, *Operations Research*, 36, 778-784.

Taylor, S.G., Seward, S.M., and Bolander, S., (1981), Why The Process Industries Are Different, *Production And Inventory Management*, 22(4), 9-24.

Tsubone, H., and Sugawara, M., (1987), A Hierarchical Production Planning System In The Motor Industry, *OMEGA International Journal Of Management Science*, 15(2), 113-120.

Wight, O., (1981), MRP II – *Unlocking American's Productivity Potential,* Essex Junction, VT, Oliver Wight Limited Publications

Winter, R., (1990), A Database Approach To Hierarchical Materials Planning, *International Journal of Operations And Production Management,* 10(2), 62-83.

Zapfel, G. (1996), Production Planning In The Case Of Uncertain Individual Demand – Extension For An MRP II Concept, *International Journal Of Production Economics*, 46-47 (Dec), 153-164.

Zijm, W.H.M., (1992), Hierarchical Production Planning And Multi-Echelon Inventory Management, *International Journal Of Production Economics*, 26(1-3), 257-264.

Zinn, W., Levy, M. and Bowersox, D.J., (1989), Measuring The Effect Of Inventory Centralization/Decentralization On Aggregate Safety Stock: The Square Root Of N Revisited, *Journal Of Business Logistics*, 10 (1), 1- 14.

Index

MIX
Papier aus verantwortungsvollen Quellen
Paper from responsible sources
FSC® C105338

If you have any concerns about our products,
you can contact us on
ProductSafety@springernature.com

In case Publisher is established outside the EU,
the EU authorized representative is:
Springer Nature Customer Service Center GmbH
Europaplatz 3, 69115 Heidelberg, Germany

Printed by Libri Plureos GmbH
in Hamburg, Germany